T0206224

Emergent Behavior in System of Systems Engineering

"This book compiles real-world case studies on discovering, understanding and engineering emergent behaviors in a computational environment across multiple application domains such as wargaming, biology, IoT, disaster management and space architecting. All the application domains are described through an undercurrent of System of Systems (SoS) engineering in conjunction with theoretical foundations required for engineering a Modeling and Simulation SoS capable of displaying valid emergent behavior. An excellent read and state-of-the-art in M&S of emergent behavior in complex systems!"

**—Dr. Saurabh Mittal, Department Chief Scientist,
The MITRE Corporation**

This book is the first of its kind to address real-world applications of the phenomenon of emergent behavior in real-world system of systems. It launches from the foundation of theory and basic understanding of the subject of emergent behavior as found in system of systems applications. It includes real-world examples where emergent behavior is manifested.

Each chapter addresses the following major points, which are exploratory in nature: the physical results of the presence of emergent behavior; the implications for the existence of emergent behavior; the manifestation of emergent behavior; and methods to either control emergent behavior assuming its effects are negative in nature, or capitalize on emergent behavior given its effects are positive in nature.

System of Systems Engineering

Series Editor:
Mo Jamshidi
University of Texas, San Antonio, USA

Intelligent Control Systems with an Introduction to
System of Systems Engineering
Thrishantha Nanayakkara, Ferat Sahin, and Mo Jamshidi

Netcentric System of Systems Engineering with DEVS Unified Process
Saurabh Mittal and José L. Risco Martín

Discrete-Time Inverse Optimal Control for Nonlinear Systems
Edgar N. Sanchez and Fernando Ornelas-Tellez

Systems of Systems Engineering: Principles and Applications
Mo Jamshidi

Emergent Behavior in System of Systems Engineering: Real-World
Applications
Larry B. Rainey and O. Thomas Holland

For more information about this series, please visit: https://www.routledge.com/
System-of-Systems-Engineering/book-series/CRCSYSENGEL

Emergent Behavior in System of Systems Engineering

Real-World Applications

Edited by
Larry B. Rainey and O. Thomas Holland

CRC Press
Taylor & Francis Group
Boca Raton London New York

CRC Press is an imprint of the
Taylor & Francis Group, an **informa** business

First edition published 2023
by CRC Press
6000 Broken Sound Parkway NW, Suite 300, Boca Raton, FL 33487-2742

and by CRC Press
4 Park Square, Milton Park, Abingdon, Oxon, OX14 4RN

CRC Press is an imprint of Taylor & Francis Group, LLC

© 2023 selection and editorial matter, Larry B. Rainey and O. Thomas Holland; individual chapters, the contributors

Reasonable efforts have been made to publish reliable data and information, but the author and publisher cannot assume responsibility for the validity of all materials or the consequences of their use. The authors and publishers have attempted to trace the copyright holders of all material reproduced in this publication and apologize to copyright holders if permission to publish in this form has not been obtained. If any copyright material has not been acknowledged please write and let us know so we may rectify in any future reprint.

Except as permitted under U.S. Copyright Law, no part of this book may be reprinted, reproduced, transmitted, or utilized in any form by any electronic, mechanical, or other means, now known or hereafter invented, including photocopying, microfilming, and recording, or in any information storage or retrieval system, without written permission from the publishers.

For permission to photocopy or use material electronically from this work, access www.copyright. com or contact the Copyright Clearance Center, Inc. (CCC), 222 Rosewood Drive, Danvers, MA 01923, 978-750-8400. For works that are not available on CCC please contact mpkbookspermissions@ tandf.co.uk

Trademark notice: Product or corporate names may be trademarks or registered trademarks and are used only for identification and explanation without intent to infringe.

ISBN: 978-0-367-75035-0 (hbk)
ISBN: 978-0-367-75037-4 (pbk)
ISBN: 978-1-003-16081-6 (ebk)

DOI: 10.1201/9781003160816

Typeset in Times
by codeMantra

Contents

SECTION III   Summary

Foreword

Emergence has become a central research topic in many domains. An emergent property is one observed in a system that cannot be explained by reductionism; it is unexpected. It is hard to cope with emergence without also dealing with complexity, as both terms are interwoven. Emergence does not emerge simply from the properties of the system's components but rather from their interplay. This relational and procedural view of a system leads to questions about how to detect emergence, how to take advantage of it, how to manage and govern it, how to avoid harm, and whether it is possible to engineer it.

This leads to the characteristics of the system. Simple systems are composed of relatively few components and can be understood by first principles, so they are fully explainable. Adding more and more components and relations quickly makes systems complicated. They are still fully explainable, but even experts require tools to support in reducing them to known inputs and outputs connected to each component. For the layman, who lacks access to the depth of knowledge needed to explain the properties, some of the exposed system properties seem to arise out of nowhere as their roots are deeply buried in the system. What looks like a non-predictable, emergent property to the layman is a predictable property to the expert. Purely computational systems are by definition compositions of computable functions, but they quickly become so complicated that even experts cannot grasp them in their entirety.

Complex systems add a new layer, beyond being complicated and huge. Complex systems are non-reducible. This ties the idea back to the emergent properties that arise from the interplay of components. Emergence is therefore tightly coupled to complex systems. How the parts and their interaction result in an observable emergence on the system level is the overarching challenge of complexity science.

But why should we care about these theoretic explanations of systems, complexity, and emergence? Because emergence happens in real life, and it has real consequences. When understood, we can use it to our advantage. If it is not understood, it can lead to poor decisions with harmful if not catastrophic consequences. This is what makes this book essential not only for scholars but also for practitioners, as it looks at real-world emergence and how to detect it and explain it, evaluates how to harness good and prevent bad emergence, and overall provides guidance on how to govern such systems. This includes the need to train decision-makers to understand the system in which they are operating. Critical thinking needs to be supported by providing decision-makers with experiences, whether through storytelling or immersive training. This requires the development of models by complexity and emergency experts that provide information that is understandable and usable to decision-makers.

One aspect to focus on is the understanding of autonomous, intelligent systems. The Internet of Things in conjunction with cloud and fog computing provides the computational foundation to allow for a new era of cyber-physical systems that must cooperate with each other, as well as with humans. This leads to an increase in the capabilities applicable to addressing a challenge, and this overlap of capabilities results in complexity and possible emergence, as the new interplay of capabilities gives rise to new, unexpected macro-behavior. This is not only true in the defense domain. It is already a reality in industry, where more production lines and processes are automated every day, and Industry 4.0 efforts are looking to extend these principles. This also requires thinking about better architectural support for the underlying networks, including improved reliability, and resilience against cyber-attacks. What has been an enabler in recent years has suddenly become an essential part of the system, contributing to the operational agility of the system's users.

The increasing awareness that we operate not only in complicated but complex systems should not lead to abandoning established methods and tools. Many are still useful in understanding and governing a system. Traditional methods need to be augmented, but many lessons learned regarding formalisms to capture systems and model language, statistics, and more are still applicable when the constraints are known and considered. While many principles valid for reducible, i.e., simple and complicated systems, no longer hold in complex systems, the gap between what is theoretically feasible and what is practically achievable is still huge, and emergence in real-world applications is more concerned with the latter. As such, the increasing use of computers to solve mathematical challenges describing complex systems and simulate them within the constraints of the possible allows us to address many challenges that would otherwise not be practically achievable.

This book has its roots in the volume "Modeling and Simulation Support for System of Systems Engineering Applications," which I edited with Larry Rainey in 2015. In that book, we took a first look into the interplay of system of systems, the possibility and importance of emergence, and how simulation can help to better cope with these phenomena. Our approach was soon followed by the volume "Engineering Emergence: A Modeling and Simulation Approach" edited by Larry Rainey and Mo Jamshidi in 2018, in which the contributing authors evaluate the applicability of modeling and simulation to define, specify, quantify, and document the domain space of emergence. This new book closes the circle by focusing on real-life applications of ideas and concepts that have been implemented in recent years and help to detect emergence, understand it, harvest positive and avoid negative emergence, and provide better decision support for those who are facing complex solution spaces.

In conclusion, this book is an important and timely contribution to the broader understanding of complexity and emergence. The increasing interest in various application domains is not only of theoretical interest but also necessary for understanding how to translate research insights into better methods and tools that support real-world operations. The chapters in this book show how we can produce, discover, explain, and manage emergence to educate decision-makers and users and ultimately allow the engineering of such properties in systems. This book makes the first steps in this direction, which will change our ways of coping with real-world emergence for the better.

Charlottesville, 25 June 2021
Andreas Tolk, PhD
Senior Principal Chief Scientist, The MITRE Corporation
Fellow of the Society for Modeling and Simulation

Editors

Larry B. Rainey is a retired US Air Force Research and Development officer. Following his retirement from the Air Force, he went to work for The Aerospace Corporation and various Department of Defense contractors. He has taught Spacecraft Design at the Air Force Institute of Technology, the graduate school of the Air Force and has taught modeling and simulation courses at Colorado Technical University and the University of Colorado at Colorado Springs. He is the executive editor for five texts all addressing the use of modeling and simulation in various venues. He is also the coauthor of the award-winning article "Harnessing Emergence: The Control and Design of Emergent Behavior in System of Systems Engineering" that has over 2000 mentions on Academia.edu and has published many articles on the application of cybernetics and systems theory to complex real-world venues.

O. Thomas Holland is a principal research engineer at the Georgia Tech Research Institute (GTRI), Information and Controls Laboratory. With his experience and degrees in electrical engineering and modeling and simulation, he is the technical area expert for modeling and simulation of emergent behavior systems and the development of solutions for modeling systems and managing the design, development, and demonstration of simulation experiments. Prior to joining GTRI, Dr. Holland was the senior strategist for modeling and simulation at the Naval Surface Warfare Center in Dahlgren Virginia, chief engineer for the Marine Corps Framework for Assessing Cost and Technology, and member of the Assistant Secretary of the Navy Research, Development, and Acquisition Modeling & Simulation Leadership Council. Spanning 36 years of experience, Dr. Holland holds four patents with the Department of the Navy and has authored topics in the detection and measurement of emergent phenomena, design and analysis of complex systems, model-based systems engineering, machine intelligence, computer architectures, image analysis, signal processing, automatic target recognition, industrial non-destructive testing, and the use of modeling and simulation for military effectiveness.

Contributors

Omar Al-Jarrah
Department of Computer Engineering,
    College of Computer and
    Information Technology
Jordan University of Science and
    Technology
Irbid, Jordan

Shelley P. Gallup
Naval Postgraduate School
Monterey, California

Kristin Giammarco
Department of Systems Engineering
Naval Postgraduate School
Monterey, California

O. Thomas Holland
Information and Communications
    Laboratory
Georgia Tech Research Institute
Atlanta, Georgia

Grace M. Hwang
Johns Hopkins University Applied
    Physics Laboratory (JHUAPL)
Laurel, Maryland
and
Kavli Neuroscience Discovery Institute
Johns Hopkins University
Baltimore, Maryland

Moath Jarrah
Department of Computer Engineering
    College of Computer and
    Information Technology
Jordan University of Science and
    Technology
Irbid, Jordan

Polinpapilinho F. Katina
Department of Informatics and
    Engineering Systems,
University of South Carolina Upstate
Spartanburg, South Carolina

Charles B. Keating
Department of Engineering
    Management and Systems
    Engineering,
Old Dominion University
Norfolk, Virginia

William J. Lademan
Marine Corps Warfighting Laboratory
Wargaming at US Marine Corps
Quantico, Virginia

Kent D. Lambert
BlockFrame, Inc.
Colorado Springs, Colorado

Larry B. Rainey
Integrity Systems and Solutions of
    Colorado, LLC
Colorado Springs, Colorado

Elizabeth P. Reilly
Johns Hopkins University Applied
    Physics Laboratory (JHUAPL)
Laurel, Maryland

Anshu Saksena
Johns Hopkins University Applied
    Physics Laboratory (JHUAPL)
Laurel, Maryland

Kevin Schultz
Johns Hopkins University Applied
   Physics Laboratory (JHUAPL)
Laurel, Maryland

Marisel Villafañe-Delgado
Johns Hopkins University Applied
   Physics Laboratory (JHUAPL)
Laurel, Maryland

Section I

Introduction and Overview

1 Introduction and Overview for Emergent Behavior in System of Systems
Real-World Applications

Larry B. Rainey
Integrity Systems and Solutions of Colorado, LLC

CONTENTS

1.1 INTRODUCTION AND OVERVIEW

In general, the term "emergence" is associated with the field of complexity. Yaneer Bar-Yam (2011) of the New England Complex Systems Institute has stated:

Emergence refers to the existence or formation of collective behaviors—what parts of a system do together that they would not do alone.

In describing collective behaviors, emergence refers to how collective properties arise from the properties of parts, how behavior at a larger scale arises from the detailed structure, behavior and relationships at a finer scale. For example, cells that make up a muscle display the emergent property of working together to produce the muscle's overall structure and movement. A water molecule has emergent properties that arise out of the properties of oxygen and hydrogen atoms. Many water molecules together form river flows and ocean waves. Trees, other plants and animals form a forest.

When we think about emergence we are, in our mind's eye, moving among views at different scales. We see the trees and the forest at the same time, in order to see how the trees and the forest are related to each other. We might consider particularly those details of the trees that are important in giving rise to the behavior of the forest.

In conventional views the observer considers either the trees or the forest. Those who consider the trees consider the details to be essential and do not see the patterns that arise when considering trees in the context of the forest. Those consider the forest do not see the details. When one can shift back and forth between seeing the trees and

DOI: 10.1201/9781003160816-2

the forest one also sees which aspects of the trees are relevant to the description of the forest. Understanding this relationship in general is the study of emergence.

Emergence can also describe a system's function—what the sys does by virtue of its relationship to the environment that would not do by itself.

In describing function, emergence suggest that there are properties that we associate with systems that are actually properties of the relationship between a system and its environment.

Consider a key. A description of a key's structure is not enough to show us that it can open a door. To know whether the key can open a door we need descriptions of both the structure of the key and the structure of the lock. However, we can tell someone that the function of the key is to unlock the door without providing a detailed description of either.

One of the problems in thinking about complex systems is that we often assign properties to a system that are actually properties of a relationship between the system and its environment. We do this for simplicity, because when the environment does not change, we need only describe the system, and not the environment, in order to describe the relationship. The relationship is often implicit in how we describe the system.

The concept of emergence as referring to function in an environment is related to the concept of emergence as the rise of collective behaviors, because any system can be viewed along with the parts of its environment as together forming a larger system. The collective behaviors due to the relationship of the larger system's parts reflect the relationships of the original system and its environment.

This is the world's first text to address real-world applications of the phenomenon of emergent behavior in real-world system of systems (SoS). To start with, we define emergence as the potential outcome/result (negative or positive) of the interaction of dissimilar and non-connected systems. It's not a matter of *if* emergence will occur, but rather *how* and *when*. As such, emergence, in and of itself, is not negative or positive per se. Rather, it is the registered effect upon the SoS and noted by the observer that classifies it as positive or negative. Therefore, it is observer dependent as to its classification, based upon the observer's understanding of imputed knowledge. As such, this phenomenon is classified as stochastic in nature. I posit that this phenomenon can occur in different domains where there is the intersection of dissimilar and non-connected systems. Also, it is postulated that this phenomenon can occur in different environments/disciplines (e.g., engineering, physics, computer science, etc.). As such, three basic questions arise for examination: (1) how does one detect the existence or presence of emergence? (2) If emergence is deemed to be positive, how does one capitalize on this effect? (3) If the emergence is deemed to be negative, how does one control this effect? As for the first question above, Chapter 15 edited by Jamshidi and Rainey addresses this subject. That is agent-based modeling can be applied to the applicable modeling and simulation environment to test for the presence of emergence.

With the above in mind for application in this text, the following definitions are provided for positive and negative emergence. Paraphrasing Mittal and Rainey (2015) and Zeigler (2016), the two definitions follow:

Positive emergence is that which fulfills the SoS purpose and keeps the constituent systems operational and healthy in their optimum performance ranges.

Negative emergence is that which does not fulfill the SoS purposes and manifest undesired behaviors.

Chapter 18 of the text edited by Jamshidi and Rainey also addresses positive and negative emergence. Favorable outcomes are associated with positive emergence, and unfavorable outcomes are associated with negative emergence.

A variation on the above themes is addressed by Grieves and Vickers (2017). They state System Behavior can be divided into two major categories. They are Predicted Behavior and Unpredicted Behavior. Predicted Behavior can be divided into Predicted Desirable (PD) and Predicted Undesirable (UU). Similarly, Unpredicted Behavior can be divided into Unpredicted Desirable (UD) and Unpredicted Undesirable (UU). In summary, there are two sources for desirable outcomes: predicted and unpredicted. Similarly, there are two sources for undesirable outcomes: predicted and unpredicted.

1.2 BACKGROUND FOR THE TEXT

The obvious question is what are the major articles and texts that have preceded this text? This section addresses this subject. This text deals with emergence found in man-made systems, in general, and in SoS engineering applications specifically. For the former application, the above definition applies. For the latter definition, Maier (1998) has defined emergent behavior as follows:

> The system of systems performs and carries out purposes that do not reside in any component system. The behaviors are emergent properties of the entire system of systems and not the behavior of any component system. The principal purposes supporting engineering of these systems are fulfilled by these emergent behaviors.

The reason for or cause of emergent behavior in system of systems has also been addressed by Maier (1998). He states: "Emergent properties can appear only if information exchange is sufficient." This, of course, refers to his "five principal characteristics of a SoS: (1) Operational Independence of the Elements, (2) Managerial Independence of the Elements, (3) Evolutionary Development, (4) Geographic Distribution, (5) Emergent Behavior." Emphasis is on the first four of these five principal characteristics. This author suggests that "information exchange" refers to software that facilitates (i.e., communicates) a common or single primary mission being performed among all the operational independent elements. It is this "mission software" that gives the SoS in question a single or primary purpose.

There have been two major texts that have been developed that deal with emergent behavior in complex systems and SoS. The predecessor to these two texts is *Modeling and Simulation Support for System of Systems Engineering Applications* edited by Rainey and Tolk (2015). This text provides a comprehensive overview of the underlying theory, methods, and solutions in modeling and simulation support for SoS engineering. Highlighting plentiful multidisciplinary applications of modeling and simulation, the book uniquely addresses the criteria and challenges found within the field.

The text *Emergent Behavior in Complex Systems Engineering: A Modeling and Simulation Approach* was edited by Diallo, Mittal, and Tolk (2018). In this text, the authors present the theoretical considerations and the tools required to enable the study of emergent behaviors in man-made systems. Information Technology is key to today's modern world. Scientific theories introduced in the last five decades can now be realized with the latest computational infrastructure. Modeling and simulation, along with Big Data technologies, are at the forefront of such exploration and investigation.

The text *Engineering Emergence: A Modeling and Simulation Approach* was edited by Jamshidi and Rainey (2019). This book examines the nature of emergence in the context of man-made (i.e. engineered) systems, in general, and SoS engineering applications, specifically. It investigates emergence to interrogate or explore the domain space from a modeling and simulation perspective to facilitate understanding, detection, classification, prediction, control, and visualization of the phenomenon. Written by leading international experts, the text is the first to address emergence from an engineering perspective.

It is then with the background of the above major article and the three texts that this text is the very first to address real-world applications of emergent behavior in SoS.

1.3   ORGANIZATION OF THE TEXT

This text has 11 chapters where real-world applications of emergent behavior in SoS are unpacked.

Chapter 1 is "Introductory and Overview" by Dr Larry Rainey. It provides an overview of this text and a brief explanation for each chapter.

Chapter 2 is "An Overview of System of Systems and Associated Emergent Behavior" by Dr Thomas Holland.

Chapter 3 by Dr Kristin Giammarco is entitled "Exposing and Controlling Emergent Behaviors Using Models with Human Reasoning." This chapter will have three primary objectives. First, to introduce the tool Monterey Phoenix that can be used to develop a model and simulation of the SoS under investigation. Second, to address how Monterey Phoenix can be used to examine the presence of both negative emergence and positive emergence. Finally, she will address how Monterey Phoenix can be used to delete negative emergence in the SoS such that only positive emergence remains. This is significant as negative emergence could negatively affect the performance of the SoS under investigation and positive emergence could significantly amplify the mission of the SoS under investigation.

The fourth chapter is entitled "Future War at Sea: The US Navy, Autonomous Vessels and Emergent Behaviors" and is written by Dr. Shelley Gallup. Cybernetics began to question the ideas of systems in control and out of control in first- and second-order behaviors. The Law of Requisite Variety makes it clear that control has limits. When Ashby described first- and second-order effects, he was not thinking of autonomy or intelligent system of systems though no doubt he understood the possibilities of emergent behavior. Moving forward to the present and near future, the United States Navy (USN) has become increasingly interested in autonomy and "intelligent" vessels of all kinds, i.e., air, surface, and subsurface. The first completely autonomous ship, Sea Hunter is now undergoing testing of its primary sensors to avoid other vessels in accordance with international rules of the road. Maier's definition of emergent behavior is

> The system performs functions and carries out purposes that do not reside in any component system. These behaviors are emergent properties of the entire system of systems and cannot be localized to any components system. The principal purposes of the system-of-systems are fulfilled by these behaviors.

In Sea Hunter, there are two perception systems that compare what is being sensed and use game theory and rules of the road to provide inputs to the ship's navigation for the most acceptable maneuver. Component integration is accomplished in communication between individual components. Under most circumstances, there is high confidence of appropriateness. However, as increasingly complex circumstances are encountered what is appropriate behavior begins to break down in unexpected ways. The USN anticipates the development and employment of many autonomous vessels and aircraft all seamlessly doing the work of carrying out the commander's intentions. In peacetime, this may be possible where context is relatively stable. In war, context becomes increasingly dynamic, time critical, and with inherent loss of communications and casualties. Emergence as a property of the aggregate systems of warfighting systems cannot be anticipated. Simulation employing the same perceptual engines as found in the vessels is just now starting to experiment with different contexts by looking for what is expected and unexpected, and whether emergent behavior can be forecast within some limits of confidence. It is anticipated that emergence will become a property of the combination of automated and unmanned systems as we move from phase 0 (peace) to phases 1 and 2 (warfighting). We have not been in a combat situation with human-machine teams or autonomous vessels alone, but from past experiences, e.g., the shootdown of an Iranian civilian aircraft (USS Vincennes incident) which is required study by naval officers, and lately two collisions that left two dozen sailors dead tell us something about possibilities of emergent behaviors at the low end of the technology. This chapter invokes Ashby's Law of Requisite Variety ("every good regulator of a system must be a model of that system"[1]) to assure, without fail, that emergent behavior is addressed if and when it is manifested. The "not knowing" was ignorance coupled with "what might happen." In the coming age of autonomy, and in combat conditions, ignoring by human operators must go up, along with trust in right action. But, no one can be sure. Sparse control of unmanned vessels will break down.

Chapter 5 is authored by Kent Lambert, Colonel USAF(Ret), Dr Thomas Bradley, and Mr Christopher Gorog and is entitled "Applications of Defense-in-Depth and Zero-Trust Cryptographic Products in Emergent Cybersecurity Environments." The authors address the presence of these emergent behaviors and their negative effects, we propose controls through three ecosystem-wide strategies, Defense-in-Depth (DiD), Zero-Trust Architectures (ZTA), and secure cryptographic key provisioning. This novel solution recommends enhancing those strategies with blockchain-assisted logistics management and a Model-Based Systems Engineering (MBSE) approach, supported by the Model-Based Systems Architecture Process (MBSAP). The Block Frame, Inc., Eco-Secure Provisioning™ (ESP™) Framework, which contains secure cryptographic key provisioning, and the Philos Blockchain/Distributed Ledger™

(Philos/BDL™) demonstrate synergistic solutions through indefinitely scalable, cross-industry SoS applications using modular, loosely coupled service interfaces, multifactor identification, and applied uniformity. Project managers, designers, and systems engineers seeking ecosystem-wide solutions for logistics control and organization-wide cybersecurity can benefit from these novel SoS architectures.

Chapter 6 is entitled "Emergence in Wargaming: Extracting Operational Insight from Dynamic Game Field" and is authored by Dr. Bill Lademan and Dr. Thomas Holland. This chapter addresses wargaming as a method which embraces the flux of a reality devoid of absolute certainty in order to investigate the dynamic nature of an operational problem. Wargaming rests upon an intellectual and situational fluidity which cannot be quantified and embraces too many interacting variables to produce the precision expected in a computational result. A wargame employs a game field which is a construct containing the wargame's pertinent variables and information and which consists of a coalescence of fluid variables, impossible to predict interactions, and dynamics not subject to law. Interaction of the wargame's participants with this game field permits an emergence of outcomes which propose insights and direction on the premise of human assessment. Thus, wargaming can excel in examining fluid and complex operational situations resistant to computation in which the opposing ideas and wills of the adversaries have as much to do with outcomes as the sum and array of the systems each side brings to the fight. The work included in this chapter examines the nature, principles, operation, and outcome of the emergence process in a wargame through the lens of the two pillars of modern wargaming: wargaming at the speed of thought and the maneuver of knowledge. Within this construct, emergent behavior occurs as a result of the participants' interaction with this game field. The analysis and assessment of this emergent behavior determine the wargame's outcomes and value.

Chapter 7 is written by Dr Moath Jarrah and Dr Omar Al-Jarrah and is entitled "DEVS-Based Modeling and Simulation to Reveal Emergent Behaviors of Internet of Things Devices." The Internet of Things (IoT) is a complex SoS where billions of IoT devices are connected to the Internet. IoT devices perform their normal functionalities which result in normal traffic behaviors. IoT botnets exploit the devices to launch distributed denial of service attacks which result in abnormal traffic volumes. The abnormal behaviors emerge as a result of exploiting a large number of IoT devices which results in an unprecedented volume of traffic. This chapter uses modeling and simulation to describe the normal and abnormal traffic behaviors of different types of IoT devices. The devices are modeled using a bottom-up approach in order to handle the complexity of the entire system. DEVS formalism is used to simulate the behaviors to show that modeling and simulation environments can be used to analyze complex systems and propose detection and control mechanisms based on traffic volume and density. The mechanisms are to be deployed by system administrators to avoid negative emergence.

Chapter 8 is entitled "Analyzing Emergence in Biological Neural Networks Using Graph Signal Processing" and is authored by Dr Grace Hwang, Dr Elizabeth Reilly, Mr Anshu Saksena, Dr Kevin Schultz, and Dr Marisel Villafane-Delgado. Biological neural networks offer some of the most striking and complex examples of emergence ever observed in natural or man-made systems. Individually, the behavior of a single neuron is rather simple, yet these basic building blocks are connected through synapses to form neural networks, which are capable of sophisticated capabilities such as pattern recognition and navigation. Lower-level functionality provided by a given network is combined with other networks to produce more sophisticated capabilities. These capabilities manifest emergently at two vastly different, yet interconnected time scales. At the time scale of neural dynamics, neural networks are responsible for turning noisy external stimuli and internal signals into signals capable of supporting complex computations. A key component in this process is the structure of the network, which itself forms emergently over much longer time scales based on the outputs of its constituent neurons, a process called learning. The analysis and interpretation of the behaviors of these interconnected dynamical systems of neurons should account for the network structure and the collective behavior of the network. The field of graph signal processing combines signal processing with network science to study signals defined on irregular network structures. Here, we show that graph signal processing can be a valuable tool in the analysis of emergence in biological neural networks. Beyond any purely scientific pursuits, understanding the emergence of biological neural networks directly impacts the design of more effective artificial neural networks for general machine learning and artificial intelligence tasks across domains. To this end, we show how techniques from the developing field of graph signal processing provide a useful toolbox.

Chapter 9 is entitled "Deep Horizon: Emergent Behavior in a System of Systems Disaster" and is written by Dr Polinpapilinho Katina and Dr Charles Keating. The Deepwater Horizon disaster is a well-documented oil-spill incident that resulted in 11 fatalities and is considered the largest marine oil spill in the history of the petroleum industry. The Deepwater Horizon was an oil platform classified as an ultra-deep water, dynamically positioned, and semi-submersible oil rig for offshore drilling operations. The complexity and different systems engaged to perform the oil rig operations can be considered an SoS. Interacting systems deployed for the Deepwater Horizon included, among others, pressure/drill monitoring, automated shutoff system, and cement modeling system. The disaster experienced by Deepwater Horizon resulted from emergent behavior occurring across a holistic spectrum of SoS challenges, crossing socio-technical-economic-political dimensions. In this chapter, the authors examine emergent behavior in the Deepwater Horizon disaster from the perspective of failure in system of systems that contributed to the disaster. The Deepwater Horizon disaster has been extensively examined and documented. However, it has not been examined with respect to emergent behavior of an SoS under disaster conditions. The present examination emphasizes understanding emergent behavior with implications for design, execution, and development of system of systems. Following an initial background and development of the Deepwater Horizon as an SoS, three primary objectives are pursued. First, several of the prominent emergent (behavior) events that occurred during the disaster are identified for exploration. The approach to this identification is to analyze the extensive documentation, reporting, and literature concerning the evolution of the disaster to identify exemplary candidates for further exploration of emergent behaviors. Second, the emergent behaviors are examined from a system theoretic framework to explain their genesis, development, and impact on the SoS. This examination is focused on exploring Systems Theory-based pathologies as aberrations from normal or health SoS function. These pathologies are examined as manifestations of emergent behaviors occurring across the disaster life cycle for the Deepwater Horizon. Third, the implications for better design, execution, and development of system of systems to deal with emergent behavior more effectively are suggested.

Chapter 10 is entitled "Emergent Behavior in Space Architecting and On-orbit Space Operations." In the first half of the chapter, Dr Rainey has obtained permission from Dr Roberta Ewart, Chief Scientist of the Space and Missile Center at Los Angeles Air Force Base, CA, to reprint her American Institute of Aeronautics and Astronautics (AIAA) article entitled "Space Architecting: Seeking the Darwinian Optimum Approach." In her article, she addresses the subject of emergent behavior in space architecting from an acquisition perspective. She originally gave this paper at the AIAA SPACE 2011 Conference and Exposition on 27–29 September 2011 at Long Beach, California. In the second half of the chapter, Dr Rainey addresses how emergent behavior can be manifested in actual on-orbit space operations. This chapter then spans the gamut from system acquisition all the way to actual operations.

Chapter 11 is also written by Dr Larry Rainey. It addresses Lessons Learned and the Proposed Way-Ahead for the subject of this text.

REFERENCES

Bar-Yam, Yaneer. "Concepts: Emergence." New England Complex Systems Institute, 2011. Web. 24 March 2017. http://neci.edu/guide/concepts/emergence.html.

Diallo, Saikou, Saurabh Mittal, and Andreas Tolk. *Emergent Behavior in Complex Systems Engineering: A Modeling and Simulation Approach.* New Jersey: John Wiley and Sons, 2018.

Ewart, Roberta "Space Architecting: Seeking the Darwinian Optimum Approach". Paper presented at the American Institute of Aeronautics and Astronautics (AIAA) SPACE 2011 Conference & Exposition on 27–29 September 2011 at Long Beach, California.

Grieves, Michael and John Vickers. Digital twin: Mitigating unpredictable, undesirable emergent behavior in complex systems. In *Transdisciplinary Perspectives on Complex Systems*, edited by Franz-Josef Kahlen, Shannon Flumerfelt and Anabela Alves. Cham: Springer, 2017.

Jamshidi, Mo and Larry Rainey. *Engineering Emergence: A Modeling and Simulation Approach.* CRC Press, 2019.

Maier, Mark W. "Architecting principles for systems-of-systems." *Systems Engineering* 1, no. 4 (1998): 267–284.

Mittal, Saurabh and Larry Rainey. "Harnessing Emergence: The Control and Design and Emergent Behavior in System of Systems Engineering". Paper won the Outstanding Paper Award for the Summer Simulation Conference, 2015.

Rainey, Larry B. and Andreas Tolk. *Modeling and Simulation Support System of Systems Engineering Applications.* New Jersey: John Wiley and Sons, 2015.

Zeigler, Bernard P. "A note on promoting positive emergence and managing negative emergence in systems of systems." *The Journal of Defense Modeling and Simulation: Applications, Methodology, Technology* 13, no. 1 (2016): 133–136.

2 An Overview of System of Systems and Associated Emergent Behavior

O. Thomas Holland
Georgia Tech Research Institute

CONTENTS

2.1 INTRODUCTION

If there were complete understanding in the technical sense about terms such as "system," "system of systems," "complexity," and "emergence," then there wouldn't be much use for this book. But these terms remain ill-defined and varied in large part because we are still learning what it means to be a system, a system of systems, or what we mean when we say, "a behavior is emergent." We are still learning because we are living the old axiom that the more we learn, the more we recognize that there is still more to learn. Nevertheless, we are slowly, but surely, developing a deeper understanding of how systems interact to form new systems, and where those new systems produce abilities that are more than just the linear sum of their individual capabilities. We are learning more about how all the ways individual systems can interact, and we are increasingly getting closer to a science of emergence.

In this chapter, we set the stage for the chapters that follow by presenting some key concepts and terms that may help the reader appreciate the role of emergent behavior within the context of system of systems. We present considerations to help us recognize the importance of the types of systems and the characteristics of system of systems that make them distinct from what may be called *traditional systems*. We describe how emergent behavior is actually the goal of a system of systems and how that emergent behavior is a result of the interrelationships between the *constituents*

DOI: 10.1201/9781003160816-3

that comprise it. We attempt in this chapter to offer some clarity of the language used to express how complexity arises because of these interrelationships, the nature of this complexity, and the importance of the balance between *variety* and *constraint*. Finally, this chapter introduces the importance of simulation in exploring emergence in system of systems and especially why the systems engineer must consider emergent behavior in systems design and analysis, whether the emergence is "positive" (intentional, acceptable, or beneficial) or "negative" (unintended, unanticipated, and detrimental). Some authors use a hyphenated spelling of the term system of systems, while others do not; there does not appear to be a consensus on the spelling. For the purposes of this chapter, we will use the hyphenated spelling of the term *system of systems* to emphasize that we are referring to a class of systems engineering constructs. Words that appear in italic font indicate a term that we feel is key and for which we attempt to provide some clarity of its definition.

2.2   SYSTEMS AND SYSTEM OF SYSTEMS

What is a system? What is a system of systems? Why are they different? Why does it matter? These are questions that can confuse engineers, especially those dealing with systems engineering requirements, design, and testing. Confounding the problem is the practicality that modern systems can be very *complicated*, i.e., they can have many tightly coupled *components*, and can even be *complex*, i.e., the interaction of components can be non-linear and so be the system can be very sensitive to perturbations and initial conditions. There is an unfortunate tendency to consider any group of systems working together as a system of systems; this is a misnomer. Indeed, your cell phone or your automobile is composed of many systems, but neither should be considered a system of systems in general and we will discuss why (but they can be part of a system of systems!) As we will see, there is a distinction between a *system* and a *system of systems*. We assert that this distinction can be made clear when we consider emergence, and that understanding this has important ramifications when we are concerned with the practical matters of systems engineering.

The International Council on Systems Engineering (INCOSE) defines a system as, "an arrangement of parts or elements that together exhibit behavior or meaning that the individual constituents do not" (INCOSE, 2021). In this definition, almost anything that is made of other things is a system. INCOSE does decompose system into "engineered system," "conceptual system," "physical system," etc. Still, this speaks to a taxonomic quality of what we should consider when we see the word "system." As early as 1998, Maier noted the term "system of systems" as a separate taxonomic group from "systems" occurred frequently in both the literature and in practice without having an accepted definition (Maier, 1998). Maier states that, "'System of systems', as commonly used, suggests assemblages of components that are themselves significantly complex, enough so that they may be regarded as systems and that are assembled into a larger system." Maier emphasizes the collaborative quality often associated with system of systems and so proposed two key features that can distinguish system of systems from systems in general:

1. The components of a system of systems exhibit operational independence. That is to say, if the system of systems is disassembled into its component

systems, then the component systems must be able to usefully operate independently. We might say that the components of a system of systems have identity and purpose beyond the system of which they are a part.

2. The components of a system of systems exhibit managerial independence. This means that not only can the components of a system of systems operate independently, but that they do so separately and in pursuit of their own independent purpose when they are apart from the system of systems.

Maier and others would later observe and assert that a key concept of what distinguishes what most people think of as a system (which for clarity we may refer to as a *Traditional Monolithic System* (Fisher, 2006)) from a system of systems is that the entities of the latter are autonomous, while the entities of the former are not. Fisher calls the non-autonomous entities composing traditional monolithic system *components* while the autonomous entities of a system of systems he calls *constituents*. This thinking, that there is a fundamental independence of existence regarding the constituents of a system of systems, provides for two additional characteristics:

1. A system of systems exhibits evolutionary development. The operational and managerial independence of the constituents enables their independent evolution, and this can add substantially to the complexity of the system of systems. Additionally, this evolutionary quality of system of systems is continuous, owing to a lack of system-wide coordination of evolutionary change (Fisher, 2006).

2. System of systems exhibit *emergent behavior*. Perhaps the only term more confusing than "system of systems" is "emergent behavior." We will discuss this in greater detail in the next section, but here we will simply state that the independence associated with constituents of a system of systems encourages competition as the constituents operate in ways most advantageous to their local situation. The role of the systems engineer then becomes one of determining how much local control a constituent is allowed given that it must still fulfill its commitment to the system as a whole. That is to say, how much *variety* can be permitted to a constituent to allow it to retain some autonomy (and hence robustness to local variations) while imposing sufficient *constraint* to assure the success of the system of which it is a part? It is easy to appreciate that if constituents of a system of systems are so constrained as to lose their independent decision-making, then there can be no evolutionary development, no managerial independence, and even no operational independence. In such a case of excessive constraint, the constituents of a system of systems arguably become mere components in a system.

Table 2.1 summarizes the main concepts that distinguish traditional monolithic systems from system of systems, and the following are some key (although in no way definitive) terms based on Fisher, Maier, and others (Shah, et al., 2007; Petty, 2018; Tolk, et al., 2018) that can inform our modern understanding of emergence as a key characteristic of system of systems:

System: Any interacting or interdependent group of entities that forms a unified and purposeful whole.

TABLE 2.1

Traditional Monolithic Systems Compared to System of Systems

Traditional Monolithic Systems	System of Systems
Centralized control	Independent decisions
Brittle (linear)	Adaptive (complex)
Behaviors specified	Behaviors emerge
Closed system constraints	Unbounded constraints
Hierarchical structures (components)	Autonomous decisions (constituents)
Components may have global visibility	Constituents have local visibility
Tight coupling between components	Dynamic interactions between constituents
Components are integrated	Constituents are interoperable

System of Systems: Any adaptive, emergent, scalable, and autonomous system where the behavior changes continuously in response to influence of stakeholders and is composed of systems that are themselves autonomous.

Traditional Monolithic System: A system comprised of components without autonomy and in which evolution is rarely treated as an integral aspect of the design, implementation, management, and operational processes.

Complex System: A system whose structure and behavior cannot be deduced nor inferred from the structure and behavior of its component parts.

Entity: A principal part of a system that is irreducible with respect to the system of which it is a part. In traditional monolithic systems, entities are non-autonomous and referred to as *components*. In system of systems, entities are autonomous and referred to as *constituents*.

Components: Non-autonomous entities comprising a system, especially traditional monolithic systems.

Constituents: Autonomous entities comprising a system of systems.

Adaptive: Able to adjust roles and functionality of their components, quality of service, network structure, or other architectural characteristics to fulfill continuously changing needs.

Autonomous: The ability to exercise independent action or decision-making.

Context: The external entities and conditions of a system that if altered can produce a change in the state of the system. The context is essentially the environment in which the system exists/operates and it needs to be taken into account in order to understand the system's behavior.

Emergent: Any characteristic of a system that cannot be localized to a single independently acting constituent or to a small constant number of constituents. Emergent properties arise from the cumulative effects of the local actions and neighbor interactions of many autonomous entities.

Scalable: Able to dynamically incorporate arbitrary numbers of additional components or constituents.

Stakeholder: The human constituents of a system of systems.

2.3 EMERGENCE AND SYSTEM OF SYSTEMS

Lurking in the background of system of systems is the concept of *complexity*. Here we are referring to true complexity, not simply the complicatedness of systems with many intricate interacting elements. Complexity as associated with system of systems is rooted in complexity theory and is characterized by such qualities as sensitivity to initial conditions, composition of components, *self-organization*, and of course, emergence (Petty, 2018; Bar-Yam, 2006). Self-organization can be thought of as the spontaneous appearance of large-scale organization through limited interactions among simple components. The operative term here is "spontaneous," meaning that when considered at the component level there is no direct constraint to enforce organization. Such spontaneous self-organization is ubiquitous in nature and rarely owes to a single enforcing mechanism. Traditional man-made systems do not exhibit such self-organization, certainly not in traditional monolithic systems. In fact, it is often the goal of traditional systems engineering to strive to eliminate self-organization that might arise as a system design becomes more complicated, especially where such self-organization results in undesired effects. Reducing the chance of self-organization is usually achieved by imposing constraints on the behaviors and limits on the interactions of system components. Unfortunately, such process of ever-increasing constraint tends to make systems that are uniquely tailored to a specific function but intolerant to any change in their context. Consider a mechanical wind-up watch; each gear, each spring, each component is strongly constrained to the point of precise specification to assure that when assembled the watch functions in a very precise and predictable manner. This is an example of a complicated system. It is complicated in that there are many parts that must interact in a precisely controlled manner, each part must adhere to very tight tolerances, and the loss of a single part will render the watch unable to perform its designed function. Although complicated, the watch is not complex. There is no non-linear combination of gears at work that produces variations within the context or allows a new behavior to emerge. This rigidity of function is actually a highly desirable quality of a watch and is a good example of a traditional monolithic system.

Complex systems, on the other hand, are often characterized by robustness and self-organization. Consider a hive of bees. Bees make honey by collecting nectar, partially digesting it, and depositing it in the honeycomb of the hive. Indeed, at a certain scale of observation, one in which the bees are not observable, we could say that the hive is a honey-making machine; place a hive in a nectar-rich environment, allow some time for operation, and then remove the honey. The amount and availability of honey present in the hive can be correlated with various environmental factors and mathematical models of honey production relating to time of year, geography, plant growth, etc., and can be reasonably predicted without regard to the actual agents of the honey production. If we removed even a single component (say a gear) from the watch, it would cease to be functional; however, unlike the watch, removing some of the honey-producing constituents of the hive does not abruptly stop the hive in its function. The population of bees can vary, yet the honey production continues. Because of the bees' intrinsic behavior, combined with various environmental influences, a honeycomb emerges with measurable, predictable, and robust behavior. The hive can be considered a complex system because in part of its self-organization that occurs to produce honey. It is convenient for us to refer to such systems where the emergent behavior is the primary behavior of interest as *emergent behavior systems*. In essence, all systems are emergent behavior systems for even the function of the watch relies on the interaction of individual components; that is, the behavior we associate with a timepiece is what emerges from the relationships defined between gears, springs, etc. However, we are more interested in emergent behavior systems comprised of constituents in a complex system. Complex systems can be said to operate on the edge of chaos, with dynamic relationships between constituents. INCOSE defines complex systems in part as those in which non-trivial relationships between cause and effect, usually involving feedback, contribute to multiple effects that are non-linear (INCOSE, 2021). Figure 2.1 attempts to depict the context for system of systems relative to the other kinds of systems we encounter. On the bottom left are traditional monolithic systems, which can be very complicated indeed but do not have non-linear relationships between components and do not exhibit component-level autonomy. Such systems do not exhibit emergence in the interesting sense, that is they are strictly constrained as in Holland's Type-0 emergence (Holland, 2007). As the autonomy of the entities that comprise the system increase, the system tends to increase in complexity and so there is an increased likelihood of interesting emergent phenomena. This emergence may be intended or unintended and may be positive or negative in its effect. In the case of emergence where the behaviors and the interactions of the constituents are well known, weak emergence (Bar-Yam, 2004) may prevail. Although more complex, these systems can still be successfully designed, managed, and analyzed much the same way as traditional monolithic systems. At the right top of the scale, we see highly non-linear system constituents with a high degree of autonomy. Consequently, the likelihood of an emergent phenomena is very high, what Holland refers to as Type-4 emergence or "spooky" emergence (Rainey & Tolk, 2015).

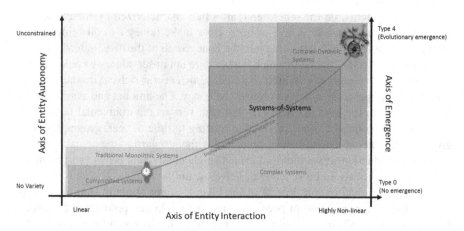

FIGURE 2.1   System of systems in the context of emergence.

Although nature capitalizes on this kind of emergence, man-made systems have tended to be limited because of our present deficit in understanding the genesis of such emergence. System of systems fall somewhere in between. They are comprised of autonomous constituents but sufficiently constrained by intent of design. Emergence is desirable for adaptability and robustness, but only enough is allowed to assure desired system behavior. High sensitivity to initial conditions and other character-istics associated with highly complex systems are avoided. It is when our designed system of systems dips its toe into the higher levels of autonomy and non-linearity that emergence that is unexpected interferes with the system of system's ability to perform as intended. This negative emergence can result from what Holland refers to as Type-4 (evolutionary) emergence in which the behaviors of system constituents change from what was previously known, from unidentified behaviors, or from unan-ticipated influences such as unanticipated variations in the environment. In general, negative emergence is to be avoided in system of systems; therefore, identifying and applying the appropriate constraining mechanisms can pose a significant challenge to the system of systems architect.

2.4 MANAGING EMERGENCE IN SYSTEM OF SYSTEMS

In 1948, Claude Shannon showed that there was a duality between energy and infor-mation (Shannon, 1948). This discovery has resulted in a measure of uncertainty for the correct transference of information between transmitters and receivers, now known as Shannon, or Information Entropy. Figure 2.2 is a diagram of Shannon's communication theory. The terms in parentheses are terms that relate Shannon's original language to a more general kind of system of information (or energy) trans-fer that can show a path from cause to effect.

Information entropy becomes particularly salient in regards to the study of systems where emergence is important as emergence in general arises from the interrelation-ship between entities. Those interrelationships can be thought of as a kind of com-munication. This communication occurs sometimes directly between entities and at other times indirectly by the entities modifying their local environment. This trans-lates to the "Channel" in Shannon's theory. This channel can take on many forms. For example, the channel can be the very environment in which entities are situated.

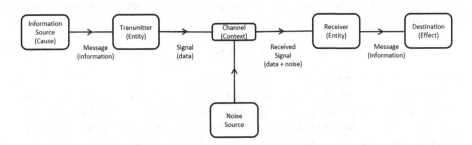

FIGURE 2.2 Shannon's general communication system.

When entities communicate through the environment, it is called stigmergy. See Grasse (1959), Theraulaz and Bonabeau (1999), and Chira et al. (2007) for insight into the biological foundation of this concept.

Shannon's theory does not rely upon a particular means of communication aside from the conceptual model of there being a transmitter, a receiver, noise source, and a channel by which communication can occur. Clearly, Shannon had electronic tele-communications in mind when he developed his theory, but we can easily see how this applies to systems of various kinds. Any device from which data can be emitted may be considered a transmitter. This could include objects such as a larynx which produces voice, a projector for images, a particle beam weapon, or even a cannon which fires a projectile. Even one particle colliding with another is a form of communication that can be interpreted in light of Shannon's theory. The important feature is that it is some item that is emitted from an entity and that it can be detected by a receiver. The receiver can be another radio device, an ear, or even the surface of an aircraft. The key here is that the receiver can accept the "information" emitted by the transmitter.

Noise in the system can be thermally induced static, weather phenomena, or active communication jamming; in short, any influence that results in the degradation of the originally transmitted information as it makes its way to the receiver. The idea of a channel is the medium in which the information can be transferred. This may be space, air, water, copper wire, etc. In radio and audio communications, we may refer to this as bandwidth. In other applications, like a road, we may refer to capacity; or in electrical distribution, we may refer to load. The important quality of the channel is that there is a fundamental limit to the amount of information that can be present in the channel. Shannon showed how each of these mechanisms, transmitter, receiver, noise, and channel are related, and his measure provides a means to quantify the uncertainty that a transmitted information will be correctly received.

Entropy (the Second Law of Thermodynamics) is typically considered a measure of heat but is rather more specifically a measure of disorder in a system and observes that closed systems progress to disorder over time. Fundamentally, entropy S is defined in terms of temperature T and heat Q by:

$$\Delta S = \frac{\Delta Q}{T} \tag{2.1}$$

This is called the *macro definition of entropy* commonly used in physics and chemistry.

Statistical mechanics takes a micro-view of physics and relates the macro definition of entropy with the number of microscopically defined states Ω accessible to a system, that is

$$S = k \ln \Omega \tag{2.2}$$

where k is Boltzmann's constant (1.4×10^{16} erg/deg).

The Second Law of Thermodynamics might seem counter-intuitive to the observation of emergent phenomena that appear to become more organized as time progresses; in fact, we observe natural emergent behavior systems organizing with great efficiency. However, this is achieved without violation of the Second Law because of the coupling between the macro levels of the system with the disorganizing process at the microlevels. H. Van Parunak referred to this as an "entropy leak" that drains disorder away from the macrolevel to the microlevel and observed that insect colonies leak entropy by depositing pheromones whose molecules evaporate and spread through the environment under Brownian motion (Parunak, 1997).

Shannon's formulation of the Second Law considers the rate at which information is produced and by taking a statistical mechanics approach considers a set of possible events $p_1, p_2, ..., p_n$. The question to be asked is, "How much choice is involved in the selection of an event?" or, rather, "How uncertain is the outcome?" Shannon showed that if there is such a measure $H(p_1, p_2, ..., p_n)$, then it must have the following properties:

1. H should be continuous in the pi.
2. Uncertainty should increase with equally likely events as the number of events increase. That is, if all the pi are equal, then $p_i = \dfrac{1}{n}$ and H should be a monotonically increasing function of n.
3. In addition, if H is decomposed, the result is a weighted sum of the decomposition of H.

Shannon showed that the only H that can satisfy all three properties is of the form

$$H(x) = -K \sum_{i=0}^{N-1} p_i \log p_i \qquad (2.3)$$

where x is a chance variable and K is a positive constant which amounts to a selection of units of measure. From the similarity of his result with the measure of entropy in statistical mechanics, Shannon referred to this as the entropy of the set of probabilities and which we now call *Information Entropy or Shannon Entropy*.

Such a relationship between entities, their complexity, and system measures was pursued by Parunak and Brueckner who observed that the Shannon Entropy of a multi-agent system is a measure of coordination of agents within a system (Parunak & Brueckner, 2001). If we consider the case of entities that can only be in one state at a time, then p_i in Shannon's entropy equation is the probability that entity i is in a specific state.

The duality between information and energy is striking, hence the term Information (or Shannon) Entropy. The information shared among entities should be of great interest to the systems engineer because what the entities in the system do in response to that information has a great deal to do with the overall system behavior. Exploring the nature of controllability in a complex system, Ashby noted that within any system, there is a possible variety of processes, but it is by the selective constraining of that variety by which useful results are obtained (Ashby, 1956, 1958). Ashby's Law of Requisite Variety relates the complexity of an overall system to the complexity of the entities comprising that system by considering the intensity of constraint defined as the ratio of the variety of the constrained system to the variety available to the system. Holland would later show that the intensity of constraint of N identical entities is given by,

$$I(t) = \frac{V_p}{V_s(t)} - 1 = \frac{N \log_2 m}{\log_2 M(t)} - 1 \tag{2.4}$$

where m is the number of unique states possible, $N \log_2 m$ is the variety available to the system if it was unconstrained, and $M(t)$ is the number of unique states exhibited by the constrained system at time t (Holland, 2007, 2018).

Statistical mechanics relates states of individual entities to overall system states. The Law of Requisite Variety similarly provides a means to express the constraint required on the entities of a system to achieve system performance. The two together, information entropy and the intensity of constraint, suggest an underlying phenomenon of relationship critical to system behavior that is manifested by observable, i.e., measurable, self-organization.

2.5 THE ROLE OF MODELING AND SIMULATION

Emergence is often a debatable concept, and some question if it really exists (Faith, 1998). Algorithmic theory and model theory deal with questions of what can really be done with computers; if all that can be produced in software is already in the rules and axioms, i.e., the program and data, can they really produce new information? Perhaps it comes down to firmly (and hopefully formally) defining what we mean by "emergence." Typically, agent-based models are used to model systems where emergence is often observed. In some cases, there is not a clear decision in the choice of modeling formalism. Discrete event simulation (DEVS) is a well-known formalism and demonstrates the property of closure under coupling (Vangheluwe, 2000), showing that coupling subsystems of a particular class produces a system of the same class. As such, it is arguable that no ontological novelty, that is some new class of system, could emerge through the interaction of the subsystems. When we model systems, it is usually to discover some previously unanticipated patterns (which we often describe as behaviors) and as such, simulation becomes an invaluable augment to our human ability to investigate and understand what might otherwise be intractably complex systems. Indeed, often more is learned about a system through the process of modeling and simulation than what is produced by the simulation itself. As George Box famously stated, "All models are wrong, but some are useful" (Box, 1979).

In spite of challenges posed concerning the theoretical underpinnings of emergence in computer simulation, progress continues to use simulations to explore systems where sophisticated entities interact in complex ways. This has value not only to aid in the engineering of system of systems but in the understanding of natural systems. In the end, it is about developing means of understanding the nature and importance of the interactions between entities within the environment in which they are situated. Understanding emergence then requires new methods to express systems and to understand the dynamics that are beyond what we might call first-order observations.

Whether in simulation or reality, what we observe as emergence is usually the evolution of a system wherein the entities comprising it create a structure by sustaining relationships; that is the interrelating entities form a new kind of entity with unique behaviors. This new entity that we observe often tends toward certain behavioral attractors, the evolution of behaviors being in response to changes in the environment, i.e., the context of the new entity. At least this gives us a mental framework for pondering what we know, what we don't know, and how we can gain insight into what we don't know from what we know. We may indeed know much about the entities we observe making up a collective, e.g., bees in a beehive. We may be able to specify the entities' behaviors in great detail, but we are often at a loss to describe the aggregate behavior of the collective in response to external influences. This is a kind of reverse reductionism, i.e., a bottom-up approach. By modeling the individual constituents of the system, we are able to explore the aggregate behavior in simulation and observe the aggregate response to various external influences. This gives us the ability to explore behaviors in the engineering sense – we can determine efficiencies and predict performance and so design to capitalize on emergence or design to avoid it. One modeling tool that begins to enable some awareness of emergent behaviors and how variety can be constrained to control emergence is MP-Firebird developed by the Naval Postgraduate School (Gimmarco, 2017). MP-Firebird has been used to show the potential for emergence in system of systems and to explore constraints to reduce unwanted emergent behaviors.

In the following chapters, we will see how emergence plays a part in different system of systems, how this emergence is managed, and how we are developing tools that can help the systems engineer understand, manage, and even design for emergence in system of systems.

REFERENCES

Ashby, W. R., 1956. *An introduction to cybernetics*. London: Chapman and Hall.

Ashby, W. R., 1958. Requisite variety and its implications for the control of complex systems. *Cybernetica*, 1(2), pp. 83–99.

Bar-Yam, Y., 2004. A mathematical theory of strong emergence using multiscale variety. *Complexity*, 9(6), pp. 15–24.

Bar-Yam, Y., 2006. Engineering complex systems: Multiscale analysis and evolutionary engineering. In: D. Braha, A. A. Minai & Y. Bar-Yam, eds. *Complex engineered systems*. Berlin, Heidelberg: Springer, pp. 22–39.

Box, G. E., 1979. All models are wrong, but some are useful. *Robustness in Statistics*, 202, p. 549.

Chira, C., Pintea, C. M. & Dumitrescu, D., 2007. Sensitive stigmergic agent systems—A hybrid approach to combinatorial optimization. In: E. Corchado, J. M. Corchado & A. Abraham, eds. *Innovations in hybrid intelligent systems*. Berlin, Heidelberg: Springer, pp. 33–39.

Faith, J., 1998. *Why gliders don't exist: Anti-reductionism and emergence*. Cambridge, MA, MIT Press.

Fisher, D. A., 2006. *An emergent perspective on interoperation in systems of systems*. Pittsburgh: Carnegie-Mellon University, Software Engineering Institute.

Giammarco, K., 2017, June. *Practical modeling concepts for engineering emergence in systems of systems*. In 2017 12th System of Systems Engineering Conference (SoSE) (pp. 1–6). IEEE. Waikoloa, Hawaii.

Grasse, P. P., 1959. La reconstruction du nid et les coordinations interindividuelles chezBellicositermes natalensis etCubitermes sp. la théorie de la stigmergie: Essai d'interprétation du comportement des termites constructeurs. *Insectes sociaux*, 6(1), pp. 41–80.

Holland, O. T., 2007. Taxonomy for the modeling and simulation of emergent behavior systems. In: *Proceedings of the 2007 Spring Simulation Multiconference-Volume 2*, pp. 28–35.

Holland, O. T., 2018. Foundations for the modeling and simulation of emergent behavior systems. In: L. B. Rainey & M. Jamshidi, eds. *Engineering emergence*. Boca Raton, FL: CRC Press, pp. 217–258.

INCOSE, 2021. *System-and SE definitions*. [Online] Available at: https://www.incose.org/about-systems-engineering/system-and-se-definition [Accessed September 2021].

Maier, M. W., 1998. Architecting principles for systems-of-systems. *Systems Engineering*, 1(4), pp. 267–284.

Parunak, H. V., 1997. "Go to the ant": Engineering principles from natural agent systems. *Annals of Operations Research*, 75, pp. 69–101.

Parunak, H. V. D. & Brueckner, S., 2001. Entropy and self-organization in multi-agent systems. In: *Proceedings of the Fifth International Conference on Autonomous Agents*, pp. 124–130.

Petty, M. D., 2018. Modeling and validation challenges for complex systems. In: L. B. Rainey & M. Jamshidi, eds. *Engineering emergence: A modeling and simulation approach*. Boca Raton: CRC Press, pp. 199–216.

Rainey, L. B. & Tolk, A., 2015. *Modeling and simulation support for system of systems engineering applications*. Hoboken, NJ: John Wiley and Sons.

Shah, N. B., Hastings, D. E. & Rhodes, D. H., 2007. Systems of systems and emergent system context. In: *5th Conference on Systems Engineering Research*.

Shannon, C. E., 1948. A mathematical theory of communication. *The Bell System Technical Journal*, 27(3), pp. 379–423.

Theraulaz, G. & Bonabeau, E., 1999. A brief history of stigmergy. *Artificial Life*, 5(2), pp. 97–116.

Tolk, A., Koehler, M. T. & Norman, M. D., 2018. *Epistemological constraints when evaluating ontological emergence with computational complex adaptive systems*. Springer: Anchorage, AK.

Vangheluwe, H., 2000. "DEVS as a common denominator for multi-formalism hybrid systems modelling." In *Cacsd. conference proceedings. IEEE international symposium on computer-aided control system design (cat. no. 00th8537)*, IEEE, pp. 129–134.

3 Exposing and Controlling Emergent Behaviors Using Models with Human Reasoning

Kristin Giammarco
Naval Postgraduate School

CONTENTS

DOI: 10.1201/9781003160816-4

3.1   INTRODUCTION

This chapter provides system of systems (SoS) architects and design practitioners with a general methodology for exposing and controlling unexpected and unwanted behaviors that could be permitted in their systems, with the objective of having only expected and acceptable behaviors permitted in a final specification. This emergent behavior analysis activity employs human reasoning together with automated tools to provide a new means for system designers to leverage positive (or acceptable) emergent behavior and remove or minimize negative (or unwanted) emergent behavior from their designs.

The enabling technology behind this methodology is Monterey Phoenix (MP) – a language, approach, and open-source tool developed by the Navy for modeling and simulating system and process behavior. MP enables the generation of sets of event traces that are orders of magnitude larger than those typically generated or manually conceived. The lightweight formal methods employed by MP furthermore guarantee completeness of the generated set of event traces up to a user-defined scope. This scope-complete set of event traces is generated more quickly and with less human error compared with manual trace generation. MP helps its users reason about intended behaviors, realize their own assumptions, turn assumptions into formal requirements, and expose and control emergent system or process behaviors. It supports modeling and simulation of SoS across many application domains and exposure and control of associated emergent SoS behaviors. A technical background is not required to become a proficient MP user, making MP a candidate for widespread use by designers of complex systems and processes in product and service industries ranging from sales and customer experience to supply chains, defense applications, and critical infrastructure.

This chapter refers to positive (desired) and neutral (allowable) emergent behaviors interchangeably with the term *acceptable*, and to the negative (undesired) emergent behaviors interchangeably with the term *unwanted*, from the point of view of the designer. Emergent behaviors may also be planned for ahead of time (*expected*) or discovered during the process of modeling (*unexpected*). For SoS modeling, emergent behaviors are those that arise in the global SoS model from the expression of individual behaviors from separate system models. The property of emergence is independent of any given human's expectation for it, which is relative (Rainey 2018).

Consider the following scenarios. A broker agent for a block chain is attacked during its construction of the initial block, and instead of halting the construction it sent the post-attack block on to the next step in the chain (Chapter 5). An Internet of Things (IoT) device sends network traffic to a server, unaware that it has been exploited by a hacker and it is part of a flooding campaign attack on a target (Chapter 7). Another IoT device goes into a periodic synchronization activity before reporting instantaneous critical sensing information, unaware that it is going to run out of battery during the synchronization activity and the sensed data will not be reported (Chapter 7). These and other scenarios were discovered in behaviors models of real systems, prompting their model authors to consider unexpected and/or unwanted behaviors latent in their designs.

3.2 MINDSET

To establish the mindset needed for understanding and employing this methodology, consider the following excerpt from a story about a sculptor explaining to a boy how he works (Pentecost 1883):

> Finally, I asked him: "Mr. M., what are you going to make out of that?" Looking up kindly into my face, he said: "My boy, I am not going to make anything out of it. I am going to find something in it." I did not quite comprehend, but said: "Why, what are you going to find in it?" He replied: "There is a beautiful angel in that block of marble, and I am going to find it. All I have to do is to knock off the outside pieces of marble, and be very careful not to cut into the angel with my chisel. In a month or so you will see how beautiful it is."

The idea can be summarized more compactly (Scott 1963):

> How do you make a statue of an elephant? Get the biggest granite block you can find and chip away everything that doesn't look like an elephant.

In both excerpts, the sculptor starts with a large block of material and then removes the unwanted material until the desired shape emerges. This same approach can be taken by practitioners to identify and remove unwanted emergent behaviors in SoS models. A model author begins with a large superset of behavior scenarios and then "chips away" all the behaviors that do not look like the wanted SoS. The methodology presented in the next section leverages development of a data set of scenarios that is the *least restrictive* to admit many possible behaviors, wanted or not. The aspect of this methodology that is typically most difficult for experienced practitioners to embrace is the necessary departure from some deeply rooted design habits that have otherwise served us well; habits cultivated with bias to describing desired interactions so well that no room remains for anything but what we know to be correct and desired to emerge in simulation. If our models start and finish as pristine images of what we want our systems to do and be, then we have no chance at using them to find and fix unwanted emergent behaviors. But if our models start as representations of known wanted behaviors along with invited or permitted misbehaviors, we have the raw material needed to temper a set of requirements that is believed to capture all expectations. This methodology entails the use of models started with excess behaviors, so that each may be finished with just the subset of behaviors its authors want to keep. The central premise of the approach is that better-behaved systems will be carved out of excess, like the angel in the marble.

3.3 EMERGENT BEHAVIOR ANALYSIS METHODOLOGY

From the point of view of this mindset, the goal of a design practitioner with access to a superset of excess behaviors is to carve away the unwanted (negative) emergent behavior and leave behind only the acceptable (positive and neutral) emergent behavior. A further goal is to remove the element of surprise: strong (unexpected) emergent behaviors that are uncovered in the process are documented and considered, and then downgraded to weak (expected) behaviors once understood. The previously unexpected/unknown behaviors, now having entered the designer's field of awareness as expected/known behaviors, fortify risk management and mitigation efforts. The designer intentionally retains the unwanted behaviors that cannot be entirely prevented in the model to formally represent and characterize risk. This model then contains unwanted behavior scenario variants that are known risks that far outnumber those describing desired behavior – the latter being those that are typically enumerated. With these goals in mind, this section lays out a general methodology with four steps for exposing and controlling emergent behaviors, including the synthetic generation of the behavior superset upon which the whole analysis depends.

3.3.1 EMERGENT BEHAVIOR DETECTION (STEP 1)

Making emergent behaviors visible in system models is prerequisite to the first step of this process: *detection*. The following principles underpin the whole methodology and especially detection because they enable the generation of the behavior superset. This superset contains not only the desired behaviors, but also the excess behaviors that may be chipped away, like the sculptor does in the story to find – not make – the angel. Adhering to these principles sets practitioners on the path to getting more emergent behaviors to become visible in their system models; later steps in the methodology explain what to do with them once they are visible.

3.3.1.1 Start with Simple, Small, and General Models

To find emergent behaviors, we start with a familiar task of modeling the system behaviors we know about or can anticipate. A basic modeling schema is sufficient to support this step. We use Auguston's abstract concept of *event* (Figure 3.1), which he defines as a fundamental building block for behavior that can be used to represent just about any activity, action, operation, process, state, transition, condition, or those of any object having behavior (Auguston 2009a). This definition leverages abstraction to unify concepts that are typically modeled as separate concepts in other modeling schemas. Allowing "events" to abstractly represent any behavior or anything with behavior is the approach we use to create simple models that give rise to complex behaviors.

Auguston's events have two basic relations: precedence (enabling events to be put in a sequence) and inclusion (enabling events to be put in a hierarchy) (Figure 3.2). With these two basic relations, events can be composed into algorithms that describe behavior using constructs of alternate, optional, and loops of zero or more or one or more iterations (Auguston 2020). Again, events can represent activities, actions,

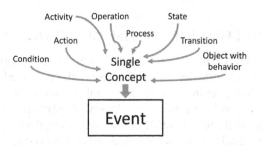

FIGURE 3.1 Auguston's concept of an abstract *event.*

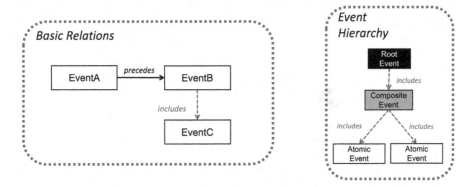

FIGURE 3.2 An event has two basic relations: precedence and inclusion.

operations, processes, states, state transitions, conditions, or any physical object with behavior, so this concept of event is flexible and permissive. Practitioners accustomed to and long satisfied with a clean distinction of separate concepts may hesitate to treat them as a single concept, but it is here argued that it is the loss of this separation that enables a gain in abstract reasoning prospects leading to discoveries that are more difficult if not impossible to make when the concepts are separated.

Concerning the content of the behavior model, a set of high-level and general events are often sufficient for producing an abundance of interesting emergent behaviors. Recalling the adage "one cannot see the forest for all the trees," it is often beneficial to model the "forest" as an event rather than the individual trees as events. Starting the model at the architecture level of design (forest) is consistent with well-known practices of stepwise model refinement and supports the discovery of certain emergent behaviors that one may not have otherwise noticed had they started with the detailed events (trees). At each level of refinement, some alternative, optional, and/or iterating events are then inserted into not just a single system of focus, but into each system member of an SoS model as described in more detail below. This multi-system behavior branching approach provides an ample supply of event combinations to inspire predictions about possible future states. To manage model scope so the model does not immediately explode when each alternative, optional, or iterating event is put in combination with every other one, the simple, small, and general model is expanded systematically from a few decision points to a large number and the model author rejects (with constraints, during the "control" step) the unwanted combinations as the expansion progresses.

3.3.1.2 Model System Behaviors and Environment Behaviors

Since much of a system's behavior depends upon what is going on around it, it is crucial to model not only the system's behaviors but also the system's environment behaviors. For SoS, this means modeling the behaviors of each system in the SoS separately from one another, and the behaviors of the SoS environment. The SoS "environment" has events composing a behavior of its own, provoking actions and responses within events composing system behavior. Behaviors in the environment also involve alternative events, optional events, and/or iterating events, and these should be described.

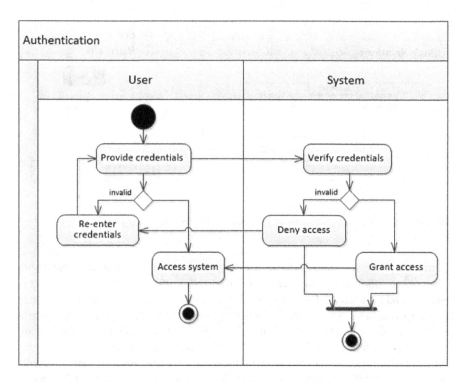

FIGURE 3.3 An example model of a system (authentication system) interacting with its environment (user) – two coordinated alternative branches in each actor.

The objective is to gain as complete a picture as possible of potential behaviors in each system and in the environment since there are often dependencies among events that go unnoticed until testing or operation. Including an environment behavior model with alternative, optional, and/or iterating event compositions primes an analyst to expose many more interaction issues that are otherwise hard to realize in advance are possible, often because of tacit assumptions that an unattended behavior cannot or will not happen, or lack of experience or awareness of examples of how an unattended behavior could happen. Figure 3.3 illustrates an example of a system interacting with its environment as a Systems Modeling Language (SysML) activity diagram of a simple authentication scenario. Specifically, a user (in the environment) provides login credentials that are then verified by the system; if the credentials are valid then the system grants access, otherwise the system denies access.

3.3.1.3 Model Interactions Separately from Behaviors and Apply Them in the Least Restrictive Way

If a model's production of desired behaviors outnumbers its production of undesired behaviors, that model may be biased toward enumeration of desired behaviors for the system (an optimism bias). Creating a formal description of stakeholder knowledge about what they need or want makes it relatively straightforward using existing methods. Experience shows, however, that there are far more ways for things to go wrong than to go right, for well-intentioned designs to be misused, attacked, disrupted, damaged, or destroyed because the stakeholders or designers were not aware of the threat at the time they needed that awareness to harden the system against it. Hindsight bias makes it tempting to take a retrospective position of "they/we should have realized that could happen – how did they/we miss that?" In reality, it is a much harder task to enumerate undesired system behaviors before they happen than to enumerate the desired system behaviors before they happen, if only because of the much higher awareness stakeholders have of the latter, so it is no wonder that most system models have an optimism bias: they pay most attention to behaviors known and stated as required. Approaches for enumerating desired behaviors, usually a finite set, do not work for enumerating undesired behaviors, a potentially infinite set which may be mostly unknown. The experts are fully aware that there are other possible behaviors, but reason that there is neither the time nor the resources to model more than the selected subset, and program management realizes and accepts that they cannot model everything. In contemporary practice, model authors typically do the best enumeration they can with the time and resources available to them, and finish when they have reached a point that stakeholders agree is "enough."

There are many existing methods for enumerating unwanted instances of behaviors during design, but they are largely informal and limited to the experience of the participating experts (e.g., the Delphi method, failure modes and effects analysis). Examples gathered from history, other projects, and personal experience make up a comparatively small set of the possible behaviors that could manifest in systems of ever-increasing complexity. Furthermore, contemporary behavior modeling techniques and tools are optimized for *describing* what the system is to do, not for *discovering* what the system should *not* be permitted to do (it is assumed the designer already knows what guard conditions and constraints are required). Notice how the interactions between the actors in Figure 3.3 are hard-coded arrows between the activities. Their presence is part of the specification, capturing that they must be there for the system to work as required. Behavior modeling approaches that treat interactions as hard-coded constraints have long been viewed as entirely acceptable. Assuming that certain interactions always take place simplifies the problem and makes a largely manual approach to behavior scenario composition manageable. Constraints appear as interactions in many popular notations and diagram formats, including Unified Modeling Language (UML) and SysML behavior diagrams, Specification and Description Language (SDL), Enhanced Functional Flow Block Diagrams (EFFBDs), Lifecycle Modeling Language (LML) Action Diagrams, and even PowerPoint-drawn sequence diagrams and flow charts. But in hard coding these interactions, could we be concealing latent possible behaviors we would prefer to expose sooner rather than later? When we specify that an interaction takes place between two events in two different systems, is that interaction then a guarantee, or is it an assumption? We may require the interaction to take place, but is it possible that under certain conditions the interaction will not take place, or will look different from how we assumed? What are the consequences? In this methodology, we probe for behavior scenarios in which interactions we assume (or require to) take place *do not actually take place* or transpire differently from our plans, such as in a different order. In most current behavior specification methods, we are submerging all behaviors except for those that we know we want, greatly impeding if not eliminating opportunities for behaviors to truly emerge in models.

A different approach is needed for undesired behavior enumeration – one that makes the excess behaviors latent in a design visible early enough in the lifecycle to impact the design. To expose and ultimately control emergent behaviors, we need to model both more and less. We need to model more alternative branches in more systems (as advised in the previous two subsections), and at the same time less forced interaction – fewer constraints than typically imposed in most contemporary behavior modeling approaches. We need more than just the known wanted behaviors in our set of scenarios; we also need to include scenarios with combinations of events that no one thought to explicitly model. For unwanted behaviors and interactions to be exposed along with desired ones, *system and environment behaviors must be modeled separately, and interactions must be modeled independent of the events they constrain*, consistent with Auguston's MP framework for system and software behavior modeling (Auguston 2020). Figure 3.4 shows this separation performed on the

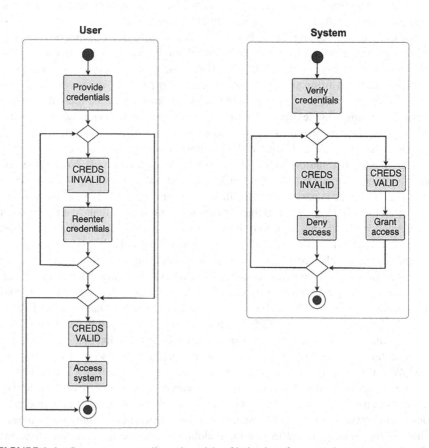

FIGURE 3.4 Separate, uncoordinated models of behaviors for an environment (user) and a system (system).

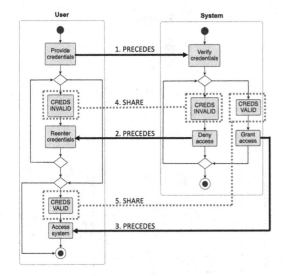

Assumed Interactions:
1. `Provide_credentials` *precedes* `Verify_credentials`.
2. `Deny_access` *precedes* `Reenter_credentials`.
3. `Grant_access` *precedes* `Access_system`.
4. `CREDS_INVALID` is the same event shared by both User and System.
5. `CREDS_VALID` is the same event shared by both User and System.

FIGURE 3.5 Interaction constraints layered over pertinent events in the actor behavior models. Solid bold lines represent precedence constraints; dashed bold lines represent event sharing (events with the same name in different actors that are the same event experienced from multiple viewpoints).

example from Figure 3.3, and Figure 3.5 shows the separation and layering on of constraints. This two-fold separation provides both the physical and mental space needed to make a fuller description of each system's behavior, while treating interactions as optional, flexible constraints that can be added or removed at will to study the effects.

3.3.1.4 Generate Behavior Scenarios from Behavior and Interaction Rules

Starting from the *least constrained* model, the generation of behavior scenarios from separately specified behavior and interaction rules enables the modeling and analysis of emergent behaviors. Figure 3.5 contains the behavior logic for the system and for the environment (at least what is known) and specifies the assumed interactions among events in different systems or parts of a system. These separate specifications provide the data set for generating behavior scenario variants with and without the interactions, which can be added or subtracted as constraints. Let us lift all constraints for an illustrative exercise. Referring to Figure 3.6, follow the behavior logic of each separate actor assuming 0 or 1 loop repetitions (we refer to the upper limit on loop iterations as the *scope*) (Auguston 2020). As the overlays show, there are four possible paths or traces through the events of the user, and there are two possible traces through the events of the system. If we put each trace in combination with every other between the two actors, assuming no constraints whatsoever, we get $4 \times 2 = 8$ traces total (Figure 3.7). Each trace is a unique combination of events from the two actors juxtaposed for inspection, assuming the absence of constraints. Comparing these traces with those generated from the activity model with hard-coded constraints in Figure 3.3, we see six more combinations. These combinations may be rejected outright as unworthy of keeping in the model because we already know (or think we know) what constraints are necessary

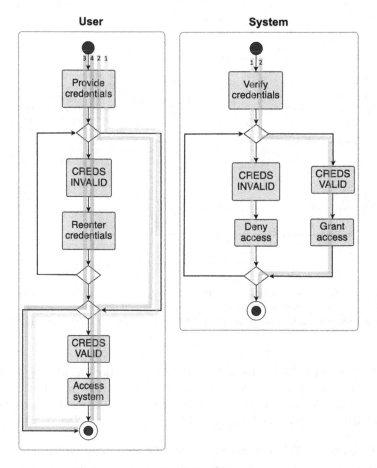

FIGURE 3.6   Trace overlays for user and system activity models showing all possible paths through each model at scope 1 (once through, with no loop repetitions). The user has four possible traces; the system has two possible traces.

for specifying the correct behavior of this simple system, but this approach shows explicitly the scenarios that we are implicitly rejecting when we hard code constraints from the start. In other words, this approach externalizes our assumptions and invites us to consider scenarios we may not have thought to consider.

Referring to Figure 3.8, follow the behavior logic of each separate actor, now assuming 0, 1, or 2 loop repetitions (scope of 2). As the overlays show, there are six possible traces through the events of the user, and six possible traces also through the events of the system. If we put each trace in combination with every other between the two actors, again assuming no constraints whatsoever, we get $6 \times 6 = 36$ traces total (Figure 3.9). The search space more than quadrupled from scope 1 to scope 2 (and it almost triples from scope 2 to scope 3 at 112 traces). This is the synthetically generated superset earlier mentioned – the block of marble obscuring the angel within it – the hunting grounds for emergent behaviors.

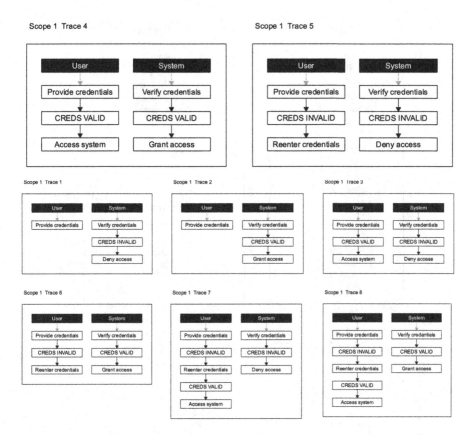

FIGURE 3.7   Event traces generated from the execution of both activity models in Figure 3.6 at scope 1, unconstrained.

Figure 3.10 shows the set of event traces resulting from the application of the constraints shown in Figure 3.5. With the constraints, there are only two traces at scope 1 and only four traces at scope 2. This constrained model represents the "angel in the marble." If we model only the angel, though, we miss out the explicit enumeration of other behaviors that might manifest without these constraints in place (e.g., all those shown in Figures 3.7 and 3.9).

When we approach a model from a *least constrained* start point, more combinations of alternative events, presence or absence of optional events, zero to several loops of events, and presence or absence of constraints are available for inspection and consideration. The graph format of the unconstrained traces is not as structured as that of a typical sequence diagram, but it paints a picture sufficient to inspire human reasoning. The number of combinations available from the systematic enumeration, even on this simple example, may exceed the number of scenarios thought of from memory or experience when drawing sequence diagrams instance-by-instance, and almost certainly will exceed it for more complex systems with many more combinations. Using the separate specifications gives us a systematic way to extract a potentially large number of scenarios that are exhaustive up to the scope but still finite for

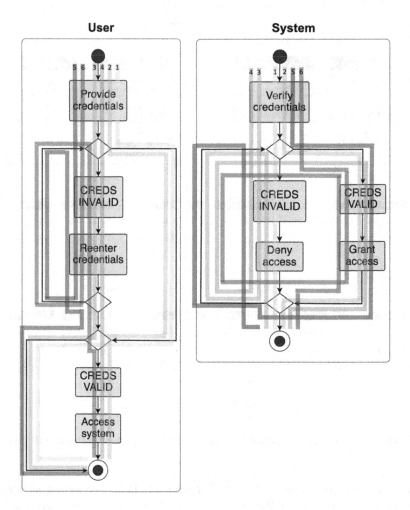

FIGURE 3.8 Trace overlays for user and system activity models showing all possible paths through each model at scope 2 (once through plus one loop repetition). The user and the system each have six possible traces.

inspection. The angel is there, but we also get the excess behaviors, which we can reject explicitly and with awareness now that we are able to see them.

3.3.1.5   Automate What You Can; Apply Cognition to Critical Inspection and Analysis

For models of simple systems, the detection step can be performed manually, as illustrated above. A skilled human is the emcee of emergent behavior modeling and analysis. The effort starts with the human cognitive effort involved in composing the behavior models that are the basis for scenario generation. The generation step illustrated in the previous section lends well to automation, conserving cognitive workload for the latter steps of the methodology that need it. Detection is good

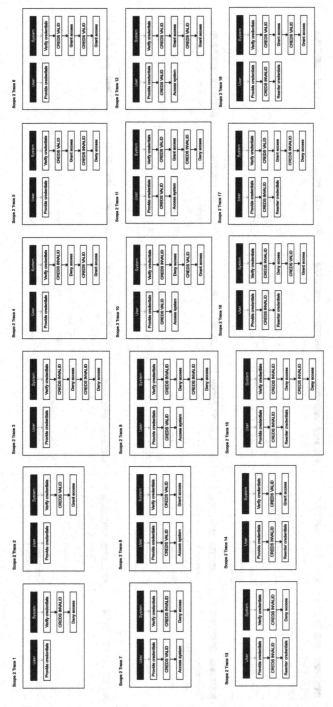

FIGURE 3.9 Thumbnail images of event traces through both activity models in Figure 3.8 at scope 2, unconstrained.

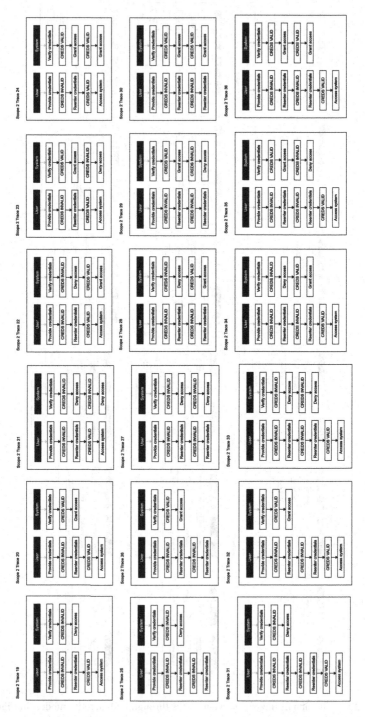

FIGURE 3.9 (*CONTINUED*) Thumbnail images of event traces through both activity models in Figure 3.8 at scope 2, unconstrained.

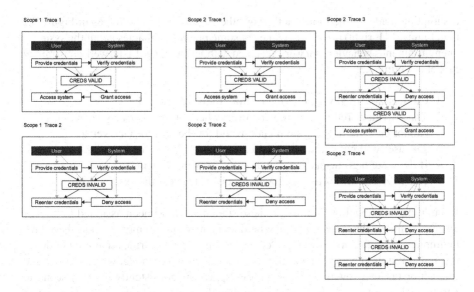

FIGURE 3.10 Interaction constraints on pertinent events from the separate actor behavior models shape the trace output, trimming away the unwanted combinations. Two traces at scope 1 are shown on the left, and four traces at scope 2 are shown on the right.

work for an automated tool because the process is systematic, can produce consistent results every time, and can scale to handle numbers of event combinations that would fatigue a human and make them prone to committing errors while manually generating a behavior superset for a substantial SoS even once, let alone redoing the generation every time something changes. If a model is small enough, then the systematic extraction of all the combinations through the events in the separate behavior specifications with or without the applied constraints could be done by hand, or even mentally. When the number of events and alternatives starts to exceed what can be processed by a person using a pencil or digital drawing, however, this step of the methodology benefits from the use of an automated tool. The first tool to do scope-complete extraction of scenarios from separately composed behavior specifications and interactions is MP-Firebird (https://firebird.nps.edu), a free and publicly available tool that implements MP behavior modeling (Auguston 2020). By automating the detection step (generation of the superset of scenarios), machines are used for the tasks most appropriate for machines and human cognitive load is reserved for tasks most appropriate for humans. The human cognitive capacity for inspection, critical thinking and reasoning, pattern detection, and idea generation is an essential part of the next steps of the methodology.

3.3.2 Emergent Behavior Prediction (Step 2)

Once the behavior superset has been generated, the next step starts with the human inspection of this least restricted set of event traces for possible fallacies – ideas that are false or mistaken (Merriam-Webster 2021) usually because of a bias held or assumption made by the model author or other stakeholder. The traces in Figure 3.7 (and Figure 3.11, right) provide evidence of some potential fallacies about the system's or environment's behavior that were accepted without challenge in the well-constrained model in Figure 3.3:

1. Credentials are always provided before they are verified.
2. The provided credentials are valid everywhere or invalid everywhere – they cannot be valid in the system and invalid for the user, or vice versa.
3. The user will not endlessly re-enter credentials following a denial.
4. If the system grants access, the user accesses the system.
5. If the system denies access, the user does not access the system.

In building the model in Figure 3.3, these statements have been assumed to be true always. Some of the statements may be documented requirements, and others may be implicit expectations viewed as "common sense" general rules that are not documented as requirements but captured as expectations in the model with a hard-coded interaction (as an arrow between two boxes appearing consistently in every scenario variant). This step of the methodology probes these assumptions and tests belief in each one's truth in all cases. The analyst(s) therefore examine the least restrictive set of event traces with intent to challenge each belief they can find. The objective, non-judgmental, under-constrained, and exhaustive generation of event combinations up to a scope during detection assists analysts in this job because it puts events in combination that are otherwise unlikely to be considered using typical methods due to the very same assumptions that result in the bias behind the omission of those combinations in the first place.

The beliefs determined from this inspection process to be potential fallacies are then the starting point for a thoughtful brainstorming activity that can be done alone or as a team activity. The participating analysts inspect the detected scenarios, challenging each belief by probing for cases in which the belief could be false. To assist this process, informal storytelling is used in which participants conceive and share conditions (single events) or stories (sequence of events) – which, if they were true in the real system or in the real environment, would provide sufficient context to explain plausible exception cases to generally accepted beliefs. The participants deploy creativity as they critically and continuously postulate potential conditions that could make each "unbelievable" event trace a valid possibility.

The main risk during the prediction step is the accidental submersion of the very behaviors that should have been allowed to emerge for consideration. Participants therefore consciously resist any temptation to complete an incomplete event trace or to "fix" the model so that an event trace looks the way they think it ought to look, because this corrective action, taken at this point in the process, impedes consideration of event traces separate from beliefs and assumptions about how the system will work or what the environment will be like, and only brings back a model such as the one shown in Figure 3.3, without the rigorous exposition and documentation of possible variants. Rather than immediately fixing what appears to be broken, participants temporarily suspend belief in what they already "know," then mindfully reflect on the unusual event combinations, and finally try to conceive of realistic circumstances in which the observed behavior might actually occur.

This prediction activity is largely unstructured due to the creative nature of brainstorming activities. Although no studies have been done to date that demonstrate one prediction activity process is better than any other in this context, practitioner experience to date seems to show that involving multiple participants with varied backgrounds working together is more productive for generating predictions than is working alone. This may be because identifying one's own beliefs and assumptions as such can be difficult without exposure to different perspectives that offer potential counterexamples. Having participants with varied backgrounds ensures the availability of an assortment of ideas and perspectives. Encouraging participants to share any idea, no matter how unusual it may seem, also seems to promote leads to valid scenarios that were previously not considered.

After the creative brainstorming process has generated as many ideas as possible, the products of that labor are documented. Figure 3.11 and Table 3.1 provide examples of how the ideas generated may be summarized and communicated using the authentication example model. The figure shows how to layer the prediction ideas over example traces using callouts, and the table shows a format for summarizing ideas in tabular format. The figure format is useful for communicating the essence of the idea to others, and the tabular format is useful for summarizing a collection of ideas for predicted behaviors.

The generalization of predicted behavior may occur after a specific instance, or story, occurs to the analyst. Stories are generated by conceiving general or specific

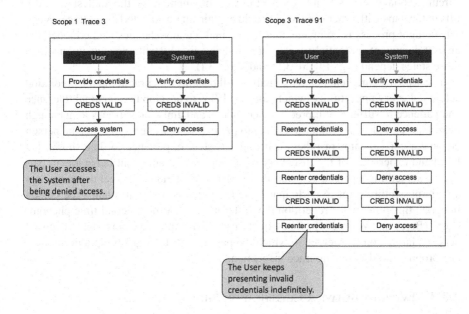

FIGURE 3.11 Prediction ideas are layered over example traces using callouts for the purpose of documenting the reviewer's thought process.

TABLE 3.1

Example of Prediction of Emergent Behaviors

Scope	Trace	Fallacy	Predicted Behaviors	Example Story
1	3	If the system denies access, the user does not access the system.	The user accesses the system after being denied access.	A person tries to gain unauthorized access to a secured area through regular access procedures but is denied. The person subsequently finds and exploits a vulnerability in the authentication protocol that enables their admission to the secured area by another means.
3	91	The user will eventually stop providing credentials.	The user keeps presenting invalid credentials indefinitely.	An unauthorized person or machine attempts authentication repeatedly in order to overwhelm system resources.

"circumstantial events" that would explain or validate the behavior seen in a questionable trace. The analyst's conception or lack of conception of circumstantial events could mean the difference between the dismissal of the scenario as a mere product of a modeling error (e.g., a misplaced event or operator symbol, a forgotten constraint) and the exposure of a valid scenario whose possibility was previously overlooked. Circumstantial events are initially held in working memory as the analyst mentally fills in the gaps with other events that could explain an unexpected event trace. When this thought process is complete, and before making any changes to the model, the analyst documents potential future states and any circumstantial events upon which those states depend (e.g., Figure 3.11 and Table 3.1).

The aspiring emergent behavior analyst will benefit from practicing the prediction activity at every opportunity, which not only helps to develop the personal tolerance and patience to suffer the imprecise, incomplete, and ambiguous traces long enough to look for and identify fallacies leading to prediction ideas but also helps the person become a good mentor of others learning this step. Experience to date shows that first-hand experience of the discovery of an unexpected emergent behavior (in particular, something a person very familiar with a system did not expect or realize was a valid possibility prior to participating in the prediction activity) helps the person integrate this process for repetition thereafter leveraging the "second time phenomenon" (Berry and Wing 1985). The extent to which unexpected emergence is exposed depends largely on the degree to which the participants are employing a permissive and probing mindset during trace inspection.

3.3.3 EMERGENT BEHAVIOR CLASSIFICATION (STEP 3)

The classification step is often done concurrently or immediately after the prediction step in practice. Classification identifies the type(s) of emergent behaviors found during detection and prediction activities. Emergent behavior may be classified as weak (expected) or strong (unexpected), and positive (desired or acceptable) or negative (unwanted or unacceptable), as defined in the following prior work. Weak emergence is "expected emergence which is desired (or at least allowed for) in the system structure" (Page 2009). Strong emergence is "unexpected emergence; that is, emergence not observed until the system is simulated or tested or, more alarmingly, until the system encounters in operation a situation that was not anticipated during design and development" (BKCASE Editorial Board 2021). Paraphrasing Mittal and Rainey (2015), Zeigler (2016) characterizes positive emergence as that which "fulfills the SoS's purpose and keeps the constituent systems operational and healthy in their optimum performance ranges" and negative emergence as that which "does not fulfill the SoS's purposes and manifests undesired behaviors." Quartuccio (2018, 2020) has used the terms "favorable" and "unfavorable" as synonyms for positive and negative respectively. Regardless of the terms used, the underlying concept and intent are the same: the permission or rejection of discovered emergent behaviors by design.

Table 3.2 develops the authentication case study to incorporate classification labels for the two example behaviors of interest.

Classification is relative to the perspective of the one(s) doing the classification. As examples, a behavior considered unexpected by a tester may be considered expected by a manufacturer, and a behavior considered positive or acceptable by a bill payer may be considered negative or unacceptable by an operator. Consider a reviewer who was, prior to seeing scope 1 trace 3, so confident in their system's security that they had not given any consideration at all to vulnerabilities that would enable such a scenario to be realized. This scenario would have surprised that reviewer. It would not surprise a reviewer who has seen such a behavior play out in a real system, so that the second reviewer may

TABLE 3.2

Example of Classification of Emergent Behaviors

Scope	Trace	Fallacy	Predicted Behaviors	Example Story	Classification
1	3	If the system denies access, the user does not access the system.	The user accesses the system after being denied access.	A person tries to gain unauthorized access to a secured area through regular access procedures but is denied. The person subsequently finds and exploits a vulnerability in the authentication protocol that enables their admission to the secured area by another means.	Strong Negative
3	91	The user will eventually stop providing credentials.	The user keeps presenting invalid credentials indefinitely.	An unauthorized person or machine attempts authentication repeatedly in order to overwhelm system resources.	Weak Negative

classify it as weak emergence. It is not so much that the combination of events is surprising, but the predictions arrived at from considering such combinations. Experts who are experienced in the subject matter but not involved in the modeling effort may be a validation source for the classification labels, but in the end, those who are involved in the modeling and design of a system day-to-day are on the front lines learning from the model. The model authors and those they have access to interact with are therefore the point of reference for the initial classification. If a prediction is surprising to them, it is of value because they are the ones who needed the awareness; if a prediction is surprising to an entire organization its value may be exponentially greater, since the top experts are learning something new about the system being modeled. Naturally, once a possible behavior and its implications have entered one's field of awareness, the surprise diminishes as the moment of realization comes and goes. A classification of strong emergence is therefore fleeting, lasting only as long as the surprise itself, but the surprising moment is preserved in order to document the learning that has occurred.

When one presents a "strong" or unexpected emergent behavior to another, hindsight bias may make the discovery appear to be more obvious in retrospect than it was at the time of realization. The perception "of course that can happen" is a common reaction to an emergent behavior found via modeling as it is to an emergent behavior found in a real system. When the Mars Climate Orbiter was lost in 1999 due to an English-to-metric conversion error (NASA 2021), it was widely criticized as a simple math error that could have easily been avoided with proper attention to detail. While it is very frustrating to learn about a simple cause to a catastrophic loss after the fact, the adage "hindsight is 20/20" applies, and the act of spotting the error in real time – by people who can do something about it at the time the action needs to be taken – may not be as easy because the participants lack the retrospective viewpoint and evidence of consequence. An unexpected behavior discovered in a model using this methodology may be initially received with similar hindsight bias. If, however, a possible behavior was not present in anyone's field of awareness in any way that made a difference for the design, and then through a model discovery activity those who needed awareness gained that awareness, then the effort was of value – possibly great value. The detection step makes example behavior instances prominently visible, unhampered by assumptions about the interactions, so that mis-combinations of events can be critically considered: "how could that happen?" It is possible (with admission that full hindsight bias is in play) that even a simple model of the Mars Climate Orbiter behavior could have produced an event mis-combination leading a reviewer to ask, "how could that happen?" without any need to address unit conversion explicitly in the model. A knowledgeable reviewer or team of reviewers may have brainstormed a prediction about unit conversion being a possible cause among other possible causes. In summary, emergent behaviors classified as "strong" are not necessarily unexpected by everyone, but perhaps just unexpected for those doing the classifying; and if they are also the ones doing the designing, then they are the ones who matter.

3.3.4 Emergent Behavior Control (Step 4)

Once emergent behaviors have been detected, predicted, and classified, the task remaining is to control these behaviors. This step involves controlling unwanted

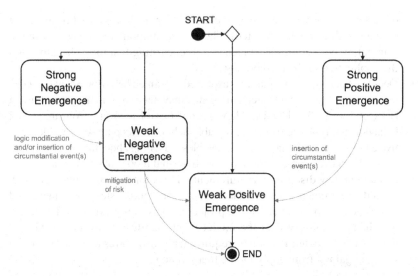

FIGURE 3.12 Progression of emergent behavior analysis. Control techniques (gray arcs) are used to downgrade strong emergence to weak emergence, and when possible, negative emergence to positive emergence.

(negative) behaviors and downgrading strong emergent behaviors to weak emergent behaviors until all that remains in the model are expected (weak) and acceptable (positive or neutral) emergent behaviors. The model progression is strong to weak, and negative to neutral and positive. At the end of the effort, the exposed unexpected and unwanted behaviors should all be purged from the design, leaving behind a well-shaped set of expected and acceptable behaviors for the SoS (Figure 3.12).

We bring unexpected emergent behaviors under control using the following procedure:

- Document any unexpected behavior immediately to preserve it before changes to the model remove it from one's field of awareness. Save a copy of the model containing the unexpected behavior(s) and annotate the model to describe the circumstantial events and predictions leading to the realization of the presence of the unexpected behavior.
- Return to the behavior model and inspect its logic. If a missing constraint or logic error was discovered (e.g., a misplaced event or operation) as the cause of unexpected emergence, fix the error(s) in a revised copy and run the model until the changes have had the expected effect on the event traces.
- Update the model with the necessary circumstantial events to document the prediction inspired by the unexpected behavior. Discern which circumstantial events need to be explicitly added as new events to the model to formalize the identified possible behavior(s). These additions should result in new alternatives and event traces that incorporate the formerly unexpected behavior, now as expected behavior.
- Downgrade the classification of the controlled unexpected, unwanted behavior. When the mechanism for the emergence is understood and the element

of surprise vanishes, the behavior classification is downgraded from strong (unexpected) to weak (expected). The documentation of its discovery in the original model remains as evidence of the learning that took place through the application of the methodology.

- Decide what to do with remaining expected unwanted behavior. Scenarios containing expected but unwanted emergence may need to be managed as known risks or further developed to show partial or full recovery from unwanted behavior to turn it into overall acceptable behavior. Keeping some weak negative emergence in a behavior model enables reasoning about failure modes and available recovery procedures along with normal modes of operation.
- Document new discovered requirements. Ideas for new or previously overlooked requirements may surface during the control step. Document and share them with members of the project so they can prevent the unacceptable behavior observed in the model from manifesting in the actual system. The documentation of new requirements also serves as evidence of the learning gained from applying the methodology.

Table 3.3 develops the authentication case study to illustrate some control strategies for the two example behaviors of interest. The need for new events for multi-factor

TABLE 3.3
Example of Control of Emergent Behaviors

Scope	Trace	Fallacy	Predicted Behaviors	Classification	Control Strategies
1	3	If the system denies access, the user does not access the system.	The user accesses the system after being denied access.	Strong Negative	Coordinate `Grant_access` with `Access_system` and coordinate `Deny_access` with `Reenter_credentials`. Insert additional authentication steps to provide additional ways to reject unauthorized users. *Discovered Requirement: The system shall employ multiple authentication factors to reject unauthorized users.*
3	91	The user will eventually stop providing credentials.	The user keeps presenting invalid credentials indefinitely.	Weak Negative	Insert a new optional event in the system called `Lock_acount` to lock the user's account after three invalid reauthentication attempts. *Discovered Requirement: The system shall lock the user's account after three invalid reauthentication attempts.*

authentication and account locking, along with some previously unstated requirements, have been exposed and documented.

Control strategies take the form of changing behavior logic (e.g., adding new alternative events) or changing or adding constraints (e.g., coordinating event interactions). Quartuccio (2020) identified four different use cases for constraints that result in the following constraint types:

- Logical constraints "limit the model from executing [logically] fallacious paths, such as receiving a message when none has been sent." For example, coordinate events that go together, so that `Grant _ access` precedes `Access _ system` and `Deny _ access` precedes `Reenter _ credentials`.
- Design constraints "limit the model from executing paths in order to control the desired pattern of the system." For example, if `Deny access` occurs three times, then `Lock _ account`.
- Simplification constraints "limit the output by restraining alternate paths that are not of interest. For example, a user may always decide to initiate a sequence, and the model may eliminate the possibility of not initiating a particular sequence." This type of constraint may be useful for controlling combinatorial explosion in larger models.
- Definition constraints "limit the output by outlining a pattern such as success-failure criteria. As an example, a correct decision within the model may [be] considered as a success criteria, and so the constraint eliminates an incorrect decision." This type of constraint may be useful also for indicating classification labels (positive or negative, favorable or unfavorable) based on rules built from known dependencies.

Figure 3.10 showed examples of the event traces with interaction constraints applied. Figure 3.13 shows an example trace with the `Lock _ account` event and constraint logic implemented.

Having now described the general concepts and methodology for emergent behavior analysis, the next section describes the language, approach, and tool that enables implementation of this analysis by providing the necessary model structure and automation of the most computationally intensive steps of the process.

3.4 MONTEREY PHOENIX

This section describes the technology enabler that makes the implementation of the emergent behavior analysis methodology practical to implement on models of real systems. Real-world applications can have complex behaviors that would make a manual analysis too labor intensive to be completed reliably, repeatably, and rapidly by those who need awareness of the emergent behaviors latent in their designs. A user-friendly approach that uses automation to do reliable, repeatable, and rapid generation of event traces is needed to make emergent behavior analysis an activity that any member of the workforce can learn and conduct in a reasonable amount of time.

Scope 3 Trace 6

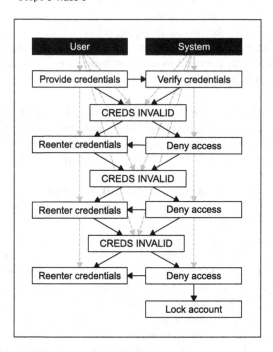

FIGURE 3.13 Example trace showing the system locking the user's account after three unsuccessful authentication attempts.

MP (Auguston 2009a, b) is a language, approach, and tool that was developed by the Navy to formally model and reason about behavior – the way in which something (e.g., a system, software, hardware, person, process) conducts activity. As a "lightweight" (Easterbrook et al. 1998, Woodcock et al. 2009) formal method and framework, it fills a capability gap that is present in other behavior modeling approaches and languages used by the Navy and other agencies in that it generates the exhaustive set of possible behavior scenarios up to a limited scope for a given model of behavior. It is also distinct from "heavyweight" formal methods in that it does not require a substantial amount of mathematical expertise to generate this exhaustive set of scenarios. Two tools, MP-Firebird (https://firebird.nps.edu) and MP-Gryphon (https://nps.edu/mp/gryphon), promote education of formal methods for system behavior modeling and make MP a cost-effective option for workforces designing complex systems at risk for exhibiting unwanted emergent behavior. MP-Firebird runs through a user's web browser requiring no installation, with models sent over an internet connection for processing. MP-Gryphon is downloaded and installed on the user's machine and models are processed locally without an internet connection. These tools are available to the public without cost or license restrictions to promote the development of safer, more secure, and more resilient systems in all industries and domains of application.

MP-Firebird and MP-Gryphon visualize the automatically generated event traces similarly in a sequence diagram like format, with MP-Firebird having a swim lane option that resembles SysML activity diagrams. Each separate actor's behavior can also be visualized as an activity diagram, and global queries can be used to extract state diagrams and component diagrams from the information in the generated set of event traces. Other capabilities of MP implemented in both tools include constraint specification and support for reasoning embracing the traditional predicate calculus notation; computations for quantities such as time, cost, and other resources using event attributes; automated, rules-based trace annotations; and reports, tables, and charts for providing summaries of properties of interest at the event trace (local) or global (across all traces) level. A collection of examples demonstrates these features and come pre-loaded with both tools available at https://nps.edu/mp/models.

3.4.1 MP Event Grammar

MP uses an abstract concept of event as the fundamental building block for behavior (Figure 3.1). An MP event has two basic relations: precedence and inclusion. Precedence is a causal dependency that is imposed to put events in a sequence, and inclusion is a parent/child dependency that is imposed to put events in a hierarchy. Using inclusion, we have three types of events in MP: root, composite, and atomic (Figure 3.2). A root event is a top-level event (has no parent) and is typically named for a top-level system activity or the after the system itself. A composite event is a mid-level event (has at least one parent and one child) and is normally used to bundle a number of related events into one event within another composite or root event. Finally, an atomic event is a leaf-level event (has no child) and represents behavior at the lowest level of granularity to be modeled.

The basic relations provide the foundation for an event grammar that is used to compose rules for behaviors and interactions within and among the modeled systems or processes. Grammar rules are formal specifications for dependencies among events. For example, the rule

```
ROOT A: B C;
```

means A is a root event that includes events B and C, with B preceding C. The colon denotes inclusion, a space or new line between events on the right-hand part of the rule denotes precedence, and a semicolon denotes the end of the rule.

Grammar rules can be used to define root events as above or composite events as follows. For example, the rule

```
B: b1 b2;
```

means that B is a composite event that includes events b1 and b2, with b1 preceding b2. The two rules together mean that A includes (B then C), where B includes (b1 then b2).

Statements written in the MP language can be viewed as a pattern to follow when generating traces. Figure 3.14 illustrates how event grammar rules with the basic structure (upper left) can be extended with syntax for other behavior patterns such as alternate (upper middle), optional (upper right), zero or more iterations (lower left),

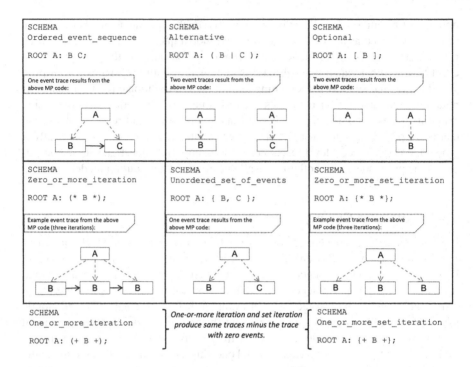

FIGURE 3.14 Example grammar rules with their corresponding event traces. (After Fig. 1 in Auguston 2020.)

one or more iterations (bottom left), unordered sets of events (lower middle), zero or more unordered iterations (lower right), and one or more unordered iterations (bottom right). When these patterns are used in combination, the right-hand part of the grammar rule builds up into an algorithm for the left-hand part. Each valid path through the algorithm may then be visualized as an event trace, which is a set of events (boxes) in terms of precedes and includes relations (arrows). We use solid arrows for precedes and dashed arrows for includes, as shown in the example event traces beneath the grammar rules in Figure 3.14. Note that one grammar rule can result in more than one event trace.

These patterns can be applied in any combination to events within a grammar rule, and rules can be nested in other rules to compose the behavior logic. Conventionally, we set up a separate grammar rule for each system or component to describe independent algorithms of behavior for each. We then use separate constraint statements to add dependencies among events in the different grammar rules. COORDINATE constraints, for example, allow one to add dependencies such as precedence, inclusion, user-defined, and other types of relations for reasoning about behavior (Auguston 2020). A named collection of related grammar rules and constraints comprises an MP schema, which is typically saved as a text file. If the user-specified rules in the schema are not contradictory or over-constrained, executing the MP schema will produce one or more valid event traces. The constraints in the schema, when crafted

```
 1    SCHEMA Authentication
 2
 3    /*-----------------------
 4     USER BEHAVIORS
 5    -----------------------*/
 6
 7    ROOT User:    Provide_credentials
 8                  (* CREDS_INVALID Reenter_credentials *)
 9                  [ CREDS_VALID Access_system ];
10
11    /*-----------------------
12     SYSTEM BEHAVIORS
13    -----------------------*/
14
15    ROOT System:       Verify_credentials
16                       (+ (   CREDS_INVALID Deny_access      |
17                              CREDS_VALID Grant_access      ) +)
18                       [ Lock_account ];
19
20    /*-----------------------
21     CONSTRAINTS
22    -----------------------*/
23
24    /* User and System share all instances of CREDS_VALID and CREDS_INVALID */
25    User, System SHARE ALL CREDS_VALID, CREDS_INVALID;
26
27    /* Instances of Provide_credentials precede instances of Verify_credentials */
28    COORDINATE  $a: Provide_credentials FROM User,
29                $b: Verify_credentials  FROM System
30        DO ADD $a PRECEDES $b; OD;
31
32    /* Instances of Deny_access precede instances of Reenter_credentials */
33    COORDINATE  $a: Deny_access        FROM System,
34                $b: Reenter_credentials FROM User
35        DO ADD $a PRECEDES $b; OD;
36
37    /* Instances of Grant_access precede instances of Access_system */
38    COORDINATE  $a: Grant_access       FROM System,
39                $b: Access_system      FROM User
40        DO ADD $a PRECEDES $b; OD;
41
42    /* Maximum number of access denials in a single session is 3 */
43    ENSURE #Deny_access <= 3;
44
45    /* Lock the account if and only if access has been denied at least three times */
46    ENSURE #Deny_access >= 3 <-> #Lock_account == 1;
47
48    /* If the account is locked, no access is granted */
49    ENSURE #Lock_account == 1 -> #Grant_access == 0;
```

FIGURE 3.15   MP schema modeling authentication. Root events are named for a user and a system with activities. A SHARE ALL statement is used to merge events in different roots, and COORDINATE statements add dependencies (PRECEDES relations in this case) between events in different roots.

with intention, prohibit invalid or unwanted combinations of behaviors from emerging in the set of generated traces.

Figures 3.7, 3.9–3.11 have shown example event traces that are produced when the schema is run using MP-Gryphon. Figure 3.15 shows the MP schema with the constraints identified for the authentication model during the prediction and control activities.

3.4.2   LEVERAGING LIGHTWEIGHT FORMAL METHODS

MP scenario generation is bound using patterns for zero-or-more or one-or-more iterations (Figure 3.14). Before executing a schema, a user assigns a maximum number of event iterations called the scope. The scope provides the upper bound on iteration at run time to produce a finite number of event traces. The critical feature that sets MP apart from other behavior modeling approaches and tools is its use of automated lightweight formal methods to generate scope-complete sets of event traces from a given schema. Scope-complete means the set of event traces is exhaustive up to the specified scope limit (Auguston 2009b, 2020); that is, every valid combination of events up to that limit is guaranteed to be present among the event traces. If an MP schema contains no iterations, the execution is exhaustive, period (it is complete).

For systems with many behavior variants, the event trace coverage provided by this automated method exceeds the coverage and speed provided by a comparable manual generation of event traces and removes opportunities for inconsistencies and errors of event trace omission. Schemas containing numerous alternatives or optional events can result in an exponential explosion of combinations as the scope increases. The best practice for managing this issue is to start with a schema with a small number of alternative or optional events and inspect resulting event traces at scope 1 before advancing to scope 2, and so on, gradually increasing the number of alternatives and scope until satisfied with the size of the set of event traces. Jackson's Small Scope Hypothesis (Jackson 2012) states that most errors can be exposed on small counterexamples; in other words, most issues that occur at high scopes can also be found at small scopes. Experience has shown that a large scope is not necessary to expose many previously unconsidered scenarios and behaviors since their shape is visible at a low scope (usually between 1 and 3). For example, in Figure 3.11 (right), three iterations are sufficient to imagine a denial-of-service attack; it is not necessary to see three hundred, thirty, or even ten repetitions to have an idea to modify the system's behavior to protect itself from too many access attempts. Three iterations are usually enough to see a pattern that would be the same, only with more repetition, at higher scopes. In this way, the Small Scope Hypothesis enables us to work in a "sweet spot" containing a finite but ample number of event traces, having scope-complete formalism and still small enough in number to allow near real-time interaction and reasoning with automated tools.

3.5   EXAMPLES

Students and faculty modeling with MP have discovered unintended, invalid, and potentially high-consequence behaviors latent within their designs fitting qualitative descriptions of weak and strong emergence. These unwanted behaviors contradicted stakeholder intent yet were not prohibited by any requirements. In this section, we present three examples of emergent behaviors found in SoS architecture models and describe them in the context of the methodology, which was developed after experience was gained from these and other examples.

3.5.1   EXAMPLES OF DETECTION

The MP-Firebird tool at https://firebird.nps.edu supports the detection of emergent behaviors by automatically generating all possible scenarios as event traces up to the scope limit. This section illustrates some detected behaviors of interest in three different models. The example models predating the current version of MP-Firebird have been re-run in the current version for uniformity in presentation. Apart from moving some boxes on the traces for better readability, the graphs have not otherwise been altered from the original student models as they were at the time of detection.

3.5.1.1 Bryant's First Responder Process

For her senior capstone project, Science and Math Academy student Jordan Bryant used MP to model a process that involves the delivery of a rescue medication called Narcan to answer a question concerning how much time is potentially saved by allowing bystanders to administer the rescue drug in comparison to waiting for the arrival of first responders (Bryant 2016). Bryant's MP model included behaviors and interactions among a victim, a bystander, and a first responder during rescue scenarios. Event traces describing typical flows showed the drug being administered either by the bystander before arrival of first responders at the scene (Figure 3.16, left) or by the first responders upon arrival. Another event trace (Figure 3.16, right) showed an alternate behavior in which the bystander called 911, then administered Narcan, then a first responder arrived and administered Narcan.

3.5.1.2 Revill's Unmanned Aerial Vehicle

For his Systems Engineering Management Master's thesis, Brant Revill used MP to model an Unmanned Aerial Vehicle (UAV) on a search and track mission for the purpose of identifying failure modes and failsafe behaviors (Revill 2016). He modeled the UAV as part of a swarm that is launched to search and track objects in an environment. A Swarm Operator controls the UAV on the mission, which may find objects of interest that turn out to be valid targets or non-targets. A critically low fuel condition (bingo fuel) was also included as a possible event. Event traces describing a typical flow showed the UAV detecting and evaluating an object in the environment, alerting the Swarm Operator who deems the object a valid target of interest, tracking the target, and finally being recalled to the base (Figure 3.17, left). Another event trace (Figure 3.17, right) showed an alternate behavior in which the UAV reached the bingo fuel condition and then subsequently found and began to track a target.

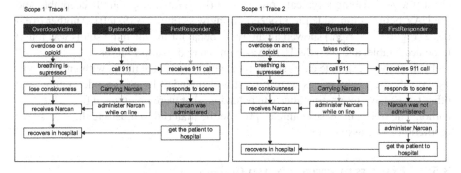

FIGURE 3.16 Model of a first responder process (Bryant 2016). An event trace (left) in which the drug was administered by a bystander instead of the first responder, and an event trace (right) in which the drug was administered by both the bystander and the first responder.

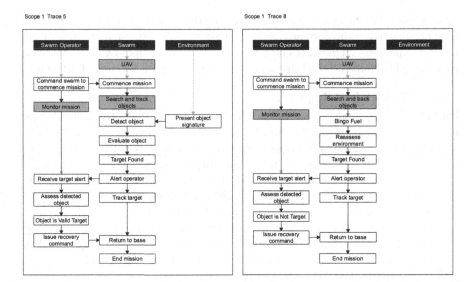

FIGURE 3.17 Model of a UAV on a search and track mission (Revill 2016). An event trace (left) in which the UAV detected and evaluated an object in the environment, tracked it, and returned to the base, and an event trace (right) in which the UAV reached a bingo fuel condition and then subsequently found and began to track a target.

3.5.1.3 Constable's UAV

Systems Engineering Management Master's student Anthony Constable used MP to model a UAV during ingress to an area in which a Humanitarian Assistance/Disaster Relief (HADR) mission is being conducted (Giammarco et al. 2018). The model involves three external systems: Joint Task Force Command and Control (JTF C2), Ground Control Station Operator (GCS Operator), and Ground Crew in addition to the UAV. In this phase, the UAV launches, climbs, and sends a status and position assessment. The phase ends either with finding the UAV status acceptable and proceeding on ingress or finding the status unacceptable and aborting the ingress. An event trace describing a typical flow showed the status being found acceptable and the UAV being commanded to proceed on ingress (Figure 3.18, left). Another event trace (Figure 3.18, right) showed an alternate behavior in which the status was found to be acceptable, and then the UAV was commanded to abort ingress.

Each of these example cases shows the detection of event traces without any analysis yet. Some of the traces were typical and intended by each model author, and other traces have interesting alternate behaviors that were unintended by each model author. The traces in these examples provide the material to illustrate the next steps: prediction and classification.

3.5.2 EXAMPLES OF PREDICTION AND CLASSIFICATION

Table 3.4 contains a summary of example fallacies, predicted behaviors, and stories generated from the detection examples along with classification assignments of weak

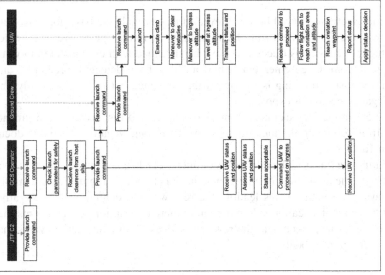
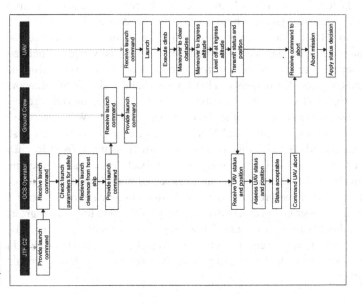

FIGURE 3.18 Model of a UAV on ingress to a HADR mission area (Giammarco et al. 2018). An event trace (left) in which the status was found acceptable, and the UAV was commanded to proceed on ingress, and an event trace (right) in which the status was found to be acceptable and then the UAV was commanded to abort ingress.

(expected), strong (unexpected), acceptable, or unwanted. A detected behavior's classification may be dependent on the circumstantial events inserted during the prediction activity, and the same event trace can lead to different predictions with different consequences.

The traces on the left side of each of Figures 3.16–3.18 are typical cases of desired behaviors for each respective system. These were documented in Table 3.4 exactly as detected without any fallacies identified or additional circumstantial events conceived during prediction. These are the scenarios that could have easily been generated manually since they are already anticipated from known design objectives, fitting the description of weak (expected) emergence. They are also valid and desired behaviors to permit in the design, so they are also classified as positive (acceptable) emergence. The emergence classification in each of these scenarios is therefore "Weak Positive."

Several cases of "Strong Positive" and "Strong Negative" emergence were also observed and documented. For example, one of the predictions made from Revill's model (scope 1 trace 8) had the UAV making it safely back to base after tracking a target in a critically low-fuel condition – an unexpected but acceptable outcome. In another example, one of the predictions made from Constable's model (scope 1 trace 2) had the GCS Operator accidentally triggering the UAV to abort ingress, sending it back to land without a Ground Crew prepared for its early return. The enumeration of combinations of events and creative storytelling over the traces has in these cases led to insights that had not been previously considered by those involved in the design work.

3.5.3 EXAMPLES OF CONTROL

The previous sections showed how MP event traces provide the basis for reasoning about each expression of behavior. This reasoning may take place over raw traces (upon detection) as well as traces that have been refined with circumstantial events (during prediction). Once emergent behaviors have been detected, predicted, and classified, the task remaining is to control the unwanted behaviors, especially any Strong Negative behaviors exposed.

Table 3.5 summarizes example control strategies and discovered requirements resulting from following the methodology for the Strong Negative behaviors documented in Table 3.4.

Annotating the traces with callouts or comments can help to convey a thought process leading to the realization of emergent behaviors to others. Figure 3.19 illustrates callouts layered over the traces leading the reviewers to identify Strong Negative emergent behavior. The callouts do not contain mere descriptions of the events that are present but instead document the reviewer's ideas for circumstantial events that are not present in the model.

3.6 CONCLUSIONS

The steps of detection, prediction, classification, and control presented in this chapter deliberately leverage human reasoning together with software automation to yield a systematic approach to exposing unexpected and/or unwanted emergent behaviors in

TABLE 3.4
Prediction and Classification Examples

Scope	Trace	Fallacy	Predicted Behaviors	Example Story	Classification
				Bryant's first responder process	
1	1	n/a	The overdose victim receives Narcan immediately from a bystander well before the arrival of first responders.	A bystander notices an opioid overdose event, calls 911, and administers Narcan medication. The first responders arrive at the scene, understand that Narcan was administered, and transport the patient to the hospital, where the patient recovers.	Weak Positive
1	2	The Overdose Victim receives Narcan from the Bystander or the First Responder, but never from both.	The first responders administer a second dose of Narcan unaware of the first dose given by the bystander.	A bystander notices an opioid overdose event, calls 911, and administers Narcan medication. The first responders arrive at the scene, think that Narcan was not administered, and administer Narcan. The patient eventually recovers despite having had double the amount Narcan believed by the hospital staff.	Strong Positive
				A bystander notices an opioid overdose event, calls 911, and administers Narcan medication. The first responders arrive at the scene, think that Narcan was not administered, and administer Narcan. The patient has a difficult recovery or does not recover due in part to the Narcan dosage misunderstanding.	Strong Negative
				Revill's UAV	
1	5	n/a	The UAV detects and tracks a valid target and completes the mission successfully.	A UAV on a mission detects and evaluates an object in the environment and begins to track it. It alerts the Swarm Operator who assesses the target's validity, the object is deemed a valid target of interest, and the UAV is called back to the base.	Weak Positive
1	−8	The UAV will not track a target in a critically low-fuel state.	The UAV detects and tracks a target while critically low on fuel but returns safely.	The UAV returns to base and makes a successful landing after tracking a target in a low-fuel condition.	Strong Positive
1			The UAV detects and tracks a target but runs out of fuel and does not return.	The UAV is forced to land at an alternative site or is lost in a crash before End_mission.	Strong Negative

(Continued)

TABLE 3.4 (Continued)
Prediction and Classification Examples

Scope	Trace	Fallacy	Predicted Behaviors	Example Story	Classification
				Constable's UAV	
1	1	n/a	The UAV uneventfully ingresses to the onstation waypoint.	A UAV launches, levels off, and transmits status and position. After the GCS Operator determines the status to be acceptable, the UAV is commanded to proceed on ingress, and the UAV continues to the onstation waypoint.	Weak Positive
1	2	The UAV reports all types of failure modes.	The UAV thinks it is in an acceptable status, but it is not.	The GCS Operator has received an indication from the Ground Crew that the system is not in an acceptable status a good report. The UAV did not detect the failure, but the Ground Crew observed a problem before it worsened, and the UAV was able to be returned to the launch site.	Strong Positive
1	2	The GCS Operator is in a good condition to work.	The GCS Operator aborts a good launch/ingress by accident.	The GCS Operator is fatigued and without realizing it, accidentally selects a command causing the UAV to return to base. Ground Crew is not aware or ready to receive the UAV.	Strong Negative
1	2	The mission will always proceed without interruption.	The mission status has suddenly changed, and the UAV is no longer needed.	All missing people being searched for in humanitarian mission have been found, and there is no longer a need to look for them.	Strong Positive

TABLE 3.5
Control Examples

Scope	Trace	Predicted Behaviors	Control Strategies	Discovered Requirement(s)
			Bryant's first responder process	
1	2	The first responders administer a second dose of Narcan unaware of the first dose given by the bystander.	Add new possible alternative events called Record_dose_amount_ and_time and Fail_to_record_dose_amount_and_time following the event Administer Narcan_while_on_line in the Bystander root. Add zero or more iteration around receives_ Narcan in the Overdose Victim root and coordinate that with possible sources of Narcan from Bystander or First Responder. Downgrade elucidated behaviors from Strong to Weak emergence.	Any Bystander who administers Narcan to an Overdose Victim shall place a band around the Overdose Victim's head that indicates the amount and time of the Narcan dose administered (Giammarco and Giles 2017).
			Revill's UAV	
1	8	The UAV detects and tracks a target but runs out of fuel and does not return.	Find and remove unnecessary redundancy in model logic. Add new circumstantial events Land_at_alternative_site and Crash to the UAV composite event as alternatives to a new event Land_at_ base. Downgrade elucidated behaviors from Strong to Weak emergence.	"If in a bingo fuel condition, the UAV shall request permission from the Swarm Operator to change course to track any new targets found." "If a possible target emerges during a bingo fuel condition, the UAV shall send the last known location of the target to the Swarm Operator and continue to return to base." "The UAV shall track targets only if it is in a non-bingo fuel condition." (Giammarco and Giles 2017).
			Constable's UAV	
1	2	The GCS Operator aborts a good launch/ingress by accident.	Place Status_acceptable event in alternate with a new event called Status_unacceptable. Add new possible events called Wrong_instruction_delivered or Misunderstanding_ of_orders preceding Command_UAV_abort in the GCS Operator root. Downgrade elucidated behaviors from Strong to Weak emergence.	GCS Operators shall have a shift duration of no longer than x hours. GCS Operators shall be trained according to standard y before operating a UAV.

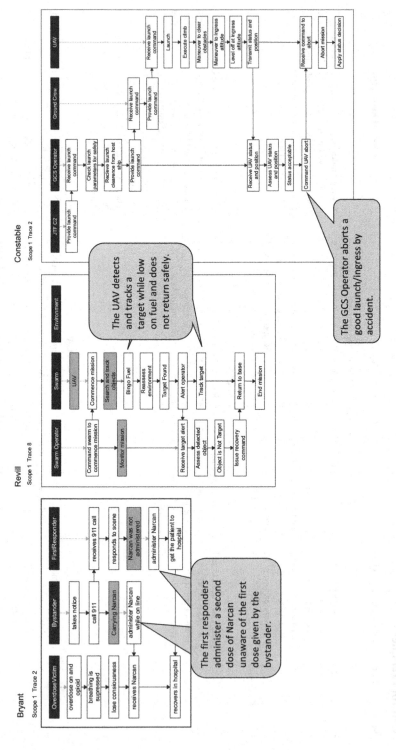

FIGURE 3.19   Prediction ideas are layered over example traces in Figures 3.16–3.18 using callouts to document ideas for new circumstantial events.

modeled systems. Human and machine tasks are interleaved to leverage the strengths of both. The detection step uses automatic MP trace generation to enumerate every combination of event flow available from a human-composed behavior model up to a given scope (number of event iterations). The ability to work with a scope-complete set of event traces is a significant advancement to emergent behavior research and practical applications since the set contains all possible paths through the behavior model, up to the scope, as separate event traces that can be inspected and queried. The prediction and classification steps make use of storytelling over the generated event traces as a means to leverage human knowledge, experience, and critical thinking. The storytelling brings new and related events to light that are not explicitly modeled ("circumstantial" events). The event traces provide inspiration to the humans reviewing them to think of circumstantial events that indicate strong emergence, or behaviors they had not previously considered or anticipated that the model would permit, along with weak emergence that they anticipated. They furthermore classify these observed behaviors as positive, neutral, or negative (or some analogous terms, like favorable/unfavorable, success/failure, etc.). Negative or unwanted behaviors are then rejected in the control step using constraints or direct changes to the system behavior logic, leaving only positive or permissible behaviors to emerge in the model. All steps together help a designer learn more about their system by prompting critical thinking about design decisions from a high-level vantage point that provides a view of the whole system of systems, both together with and separate from interactions among its parts. It is at this level of abstraction that many emergent behaviors lie dormant in designs, but they can now be exposed prior to testing or fielding by viewing the model through the lens provided by this MP-enabled methodology.

Notably, this methodology expands the sight picture for designers who are accustomed to working with relatively small subsets of weak positive or neutral and weak negative behaviors. MP enables a larger set of those types to be generated, plus peripheral strong positive or neutral and strong negative behaviors that were never part of any design process for lack of an approach to address their discovery and enumeration ahead of testing or fielding. This contribution brings formality and automation to a part of the design process that has historically not had much attention: the iterative improvement of human understanding of design problems by exposing and purging assumptions that impede the identification of extra behaviors that could either be (1) unwanted interference with successful operation of the system or (2) unrealized opportunities for better functioning of the system.

Architects working with emergent behaviors are sculptors rather than builders. A sculptor starts with an excess of material and then removes the unwanted material until the desired shape emerges. In this analogy, the excess of material are *all* instances of behavior from each separate system (unconstrained), and the unwanted material are selected behaviors that are rejected when interactions are specified as constraints. The resulting model, which includes the constraints used to shape the desired behavior, becomes a formal specification for system or SoS requirements.

Aspiring emergent behavior analysts will benefit from letting go of any compulsion to control the shape of the model from its outset embracing a flexible and permissive mindset to admit event traces they may never otherwise have constructed or considered. Irregular event traces should be tolerated long enough to consider whether they lead to reasonable and realistic predictions impacting the design of a well-behaved system. Subject matter experience brings a great advantage in familiarity with relevant systems, but also a possible pitfall of the human condition: we must approach this analysis with humility and remain cognizant of habitual thinking that could lead us to believe we already know what we want a system to do, what we want it to avoid doing, and how to achieve the outcomes we want in the design. We must, in essence, be willing to interrogate all that we think we already know and be open to considering traces that our training and experience would compel us to reject immediately (possibly due to some unrealized assumption or bias). This is where a diverse composition of team members can help expose assumptions, because not everyone makes the same ones. The way we think and reason greatly impacts the content of our models, and if we do not practice humility and questioning of our own thinking, we will most probably see in the output no more than exactly what we expected to get; and if we have made a mistake, we may fix it immediately without pondering the plausibility of the faulty trace with curiosity and learning from its implications. In the end, this is a methodology for those who are eager to improve their own thinking and reasoning using models. Automated tools support the computationally intensive parts of this activity, but it is a human, not a tool, that exposes and controls emergent behaviors.

REFERENCES

Auguston, M. 2009a. Monterey phoenix, or how to make software architecture executable. In *Proceedings of the 24th ACM SIGPLAN Conference Companion on Object Oriented Programming Systems Languages and Applications*, pp. 1031–1040.

Auguston, M. 2009b. Software architecture built from behavior models. *ACM SIGSOFT Software Engineering Notes* 34, no. 5: 1–15.

Auguston, M. 2020. *System and software architecture and workflow modeling language manual (version 4)*. Monterey, CA: Naval Postgraduate School. https://wiki.nps.edu/display/MP/Documentation (accessed November 30, 2021).

Berry, D. M. and J. M. Wing. 1985. Specifying and prototyping: Some thoughts on why they are successful. In *International Joint Conference on Theory and Practice of Software Development*, pp. 117–128. Springer: Berlin, Heidelberg.

BKCASE Editorial Board. 2021. The Guide to the Systems Engineering Body of Knowledge (SEBoK), v.1.9.1. https://www.sebokwiki.org/wiki/Emergence (accessed November 30, 2021).

Bryant, J. 2016. Using Monterey Phoenix to analyze an alternative process for administering Naloxone. Capstone research project, Science and Math Academy. http://scienceandmathacademy.com/academics/srt4/student_work/2016/bryant_jordan.pdf (accessed November 30, 2021).

Easterbrook, S., R. Lutz, R. Covington, J. Kelly, Y. Ampo, and D. Hamilton. 1998. Experiences using lightweight formal methods for requirements modeling. *IEEE Transactions on Software Engineering* 24, no. 1: 4–14.

Giammarco, K., R. Carlson, and M. Blackburn. 2018. *Verification and validation (V&V) of system behavior specifications*. Systems Engineering Research Center (SERC), Stevens Institute of Technology: Hoboken, NJ.

Giammarco, K.; Giles, K. (2017). Verification and validation of behavior models using lightweight formal methods. In *Proceedings of the 15th Annual Conference on Systems Engineering Research*, Redondo Beach, CA, 23–25 March 2017.

Jackson, D. 2012. *Software abstractions: Logic, language, and analysis*; MIT Press: Cambridge, MA.

Merriam-Webster. 2021. S.v. fallacy. https://www.merriam-webster.com/dictionary/fallacy (accessed November 30, 2021).

Mittal, S., and L. Rainey. 2015. Harnessing emergence: The control and design of emergent behavior in system of systems engineering. In *Proceedings of the Conference on Summer Computer Simulation*, pp. 1–10.

Page, S. 2009. *Understanding complexity, great courses: Business & economics.* The Teaching Company: Chantilly, VA.

Pentecost, G. F. 1883. *The angel in the marble, and other papers.* Hodder and Stoughton: London.

Planetary Science Communications Team, NASA Jet Propulsion Laboratory. n.d. Mars climate orbiter. https://solarsystem.nasa.gov/missions/mars-climate-orbiter/in-depth/ (accessed November 30, 2021).

Quartuccio, J. J., and K. M. Giammarco. 2018. A model-based approach to investigate emergent behaviors in systems of systems. In *Engineering emergence: A modeling and simulation approach*, pp. 389–458. ed. L. B. Rainey and M. Jamshidi. Boca Raton: CRC Press.

Quartuccio, J. J. 2020. Identification of behavior patterns in systems of systems architectures. PhD diss., Naval Postgraduate School: Monterey, CA, USA.

Rainey, L. B. 2018. Introduction and overview for engineering emergence: A modeling and simulation approach. In *Engineering emergence: A modeling and simulation approach*, ed. L. B. Rainey and M. Jamshidi. Boca Raton: CRC Press.

Revill, M. B. 2016. UAV swarm behavior modeling for early exposure of failure modes. Master's thesis, Naval Postgraduate School: Monterey, CA.

Scott, J. 1963. *The plain dealer, why are elephants?* Young Ohioans Editor: Cleveland, OH.

Woodcock, J., P. G. Larsen, J. Bicarregui, and J. Fitzgerald. 2009. Formal methods: Practice and experience. *ACM computing surveys (CSUR)* 41, no. 4: 1–36.

Zeigler, B. P. 2016. A note on promoting positive emergence and managing negative emergence in systems of systems. *The Journal of Defense Modeling and Simulation* 13, no. 1: 133–136.

Section II

Practical Applications

4 Future War at Sea
The US Navy, Autonomy in War at Sea and Emergent Behaviors

Shelley P. Gallup
Naval Postgraduate School

CONTENTS

Cybernetics began to question the ideas of systems in control and out of control in first and second-order behaviors. The Law of Requisite Variety makes it clear that control has limits. When Ashby described first and second-order effects, he was not thinking of autonomy or intelligent system of systems though no doubt he understood the possibilities of emergent behavior. Moving forward to the present and near future, the United States Navy (USN) has become increasingly interested in autonomy and "intelligent" vessels of all kinds, i.e., air, surface, and subsurface. An autonomous vessel and its sister ship are now undergoing testing of its primary sensors to avoid other vessels in accordance with international rules of the road and potential tactics. Maier's definition of emergent behavior is

DOI: 10.1201/9781003160816-6

The system performs functions and carries out purposes that do not reside in any component system. These behaviors are emergent properties of the entire system of systems and cannot be localized to any components system. The principal purposes of the system-of-systems are fulfilled by these behaviors.

Multiple perception systems that compare what is being sensed use game theory and rules of the road can provide inputs to the ship's navigation for the most acceptable maneuver. Component integration is accomplished in communication between individual components. Under most circumstances, there is high confidence of appropriateness. However, as increasingly complex circumstances are encountered, what is appropriate behavior begins to break down in unexpected ways. The USN anticipates the development and employment of many autonomous vessels and aircraft, all seamlessly doing the work of carrying out the commander's intentions. In peacetime, this may be possible where context is relatively stable. In war, context becomes increasingly dynamic, time critical, and with inherent loss of communications and casualties. Emergence as a property of the aggregate systems of warfighting systems cannot be anticipated. Instead of full autonomy, human to machine "teams" and "partnerships" are now also being considered. Simulation employing the same perceptual engines as found in the vessels is just now starting to experiment with different contexts by looking for what is expected and unexpected, and whether emergent behavior can be forecast within some limits of confidence. The old saying that "no plan survives first contact with the enemy" reminds us that in conflict, it is the *unexpected* that is expected, but in the space of unknown unknowns is anticipated that emergence will become a property of the combination of automated and unmanned systems as we move from phase 0 (peace) to phases 1 and 2 (warfighting). Sparse control of unmanned vessels will break down and **What** emerges as the collective strategy of unmanned platforms working within the Observe, Orient, Decide, Act (OODA) loop associated with human planners. Additionally, this chapter will consider that system of systems in military combat operations are those systems interrelated by a complex web of possible outcomes that are increasingly unpredictable. **How** it will manifest itself will be in the lack of control by human supervisors with the implication that both good and bad effects are possible (we postulate that human-machine partnerships create a greater possible emergence at a distance, and thus both need to be on the vessel). **Where** emergent behavior will manifest itself is a bit more complex. A new concept is being discussed in which the network is the warfighting platform, and ships, etc. are units within the network. In this new way of concentrating lethal power where it needs, **when** emergent behavior is most likely to show up is in the configuration of the combat network, targeting, and intentional messages supporting necessary effects.

4.1 INTRODUCTION

When you have attained the power of the long sword, you can singlehandedly prevail over ten men. When it is possible to overcome ten men singlehandedly then it is possible to overcome a thousand men with a hundred, and ten thousand men with a thousand.

(Miyamoto Musashi, The Book of Five Rings, 2006)

Our present information age illustrates the close coupling between tactics and technology and portends a sharply increased power of tactics in effecting operations and strategy. While strategy may seek to control the scope, pace and intensity of a conflict, tactics controls the very powerful second derivative, that is, the rate of change that affect men's minds where wars are won or lost. The importance of scouting (information-gaining activities) and missiles reflects the dominant characteristic of the information age—access and speed....High levels of shared awareness among forces have been shown to dramatically increase speed to scout, to decide and to execute.

(VADM Arthur K. Cebrowski, USN in forward to Fleet Tactics and
Naval Operations, 3rd edition Naval Institute Press, 2018)

On the one hand, Musashi is writing about the art of the sword and the samurai warrior's close attention to how that sword is used. This is the primary tool of the samurai, and in battle, it is clear that the more trained and greater number of swordsmen with the qualities he describes will very likely be victorious.

In the second quote, VADM Cebrowski is advancing the same thought; that the tools of war are different, moving from sword to shared awareness. It is here that the future of naval battles will be won or lost. The side which is most capable of maintaining pace with the changes in context, targeting of adversaries and accomplishing the first to fire when conflict has been entered, will emerge the victor. This is an extension of Musashi's long sword in today's world.

But, how is shared awareness accomplished? What exactly does this mean? Employing complex sensor systems and producing huge amounts of data across the force are only one part of the problem space. Data must become information, and actionable information knowledge. Knowledge of intentions, context, enemy positions, and commander's intent (whether kinetic or not) are all combined within a system of systems (SoS). People are embedded within this SoS, and what seems at first a technical challenge is now also a *socio-technical* system of interactions between humans and machines.

The term *hybrid-warfare* has become part of the military lexicon of late, meaning the incorporation of human-machine teams, or as will be discussed later in this chapter, human-machine partnerships.

This further addition to the art of naval warfare creates a need for further understanding of a four-part SoS (human to human, human to machine, machine to human, and machine to machine) makes us pause and reflect the many ways in which forces may be employed in near conflict (peace operations, but under pressure from an opponent), or at the point where conflict is engaged.

4.2 EMERGENT BEHAVIOR

What is an "emergent behavior"? Why is it important? What can we know about it, such that we are not surprised when it happens? Is it more related to the advances in systems and their complexity, or an increase in the rapidity with which people access, use, and combine these systems?

The purpose of this chapter is to drill down into the depths of maritime operational art given a near future in technical capabilities, while still asserting what has been learned about naval strategy and tactics since Thucydides documented the history of the road to war and its outcomes between Sparta and Athens in the Peloponnesian War. Great shifts in power and outcomes didn't necessarily occur in combat, but as an emergence of the arguing between powerful men in the Athenian corner. Things are not always what they seem. And in fact, after 30 years of war, neither side could solidly claim victory and both were in ruins.

It is useful now to provide a formal definition of what we will mean by an emergent behavior in this chapter. Emergence is a property in which the system performs functions and carries out purposes that do not reside in any component system. These behaviors are emergent properties of the entire SoS and cannot be localized to any component system. The principal purposes of the SoS are fulfilled by these behaviors (Maier 1998).

Two types of emergence have been defined: positive emergence and negative emergence.

Positive Emergence: That which fulfills the SoS purpose and keeps the constituent systems operational and healthy in their optimum performance ranges.
Negative Emergence: That which does not fulfill the system of systems purposes and manifest undesired behaviors (Maier 1998).

A more in depth look at SoS, their interrelations can be found in "Practical Modeling Concepts for Engineering Emergence in Systems of Systems" (Giammarco 2017).

Note that this is a distinction between behaviors that do or do not assist in meeting the intentions of the SoS being considered and should not be confused with systems with feed-forward and feed-back designs (although they can well be part of the emergent state).

Other simulations mean to try and resolve the outcomes of SoS behaviors include campaign analysis (operations analyses), Lanchester equations (ratios of factors comprising one force when comparing to another force), the salvo model (Hughes 1995), and physics-based modeling that expresses the engineering capabilities of one weapon system against another forces' are all examples. A difficulty with most of these approaches is that they require some means by which to approximate friendly and foe capabilities. Historical examples, intelligence approximations, knowledge from subject matter experts, and so forth are gathered to make the best guess possible for both sides. What is gained by these methods are courses of action that can be presented to flag officers (admirals and generals) for a final decision on deploying forces for greatest intended effect. Often these factors are incomplete, or at worst, misleading. The most famous case of this was the use of "body counts" during the Vietnam war to assess US and allied success of current strategies. A more recent example can be found in the swift and complete overtaking of Afghanistan by Taliban forces. However, for daily decision-making, where the factors are weighted properly, these approaches can produce a quick and understandable set of possible moves at the operational level. This is neither an SoS approach nor Complex Adaptive Systems of Opposing (CASOF) examination of naval forces employing a very large variety of SoS.

Alternatively, warfare at sea as a SoS of unusual size, complexity, dynamics, and results produce many possible stories of how a pending conflict will proceed, and potential for success. Many different means to war game scenarios have been tried ahead of time. War games (e.g., plan Orange between WW 1 and WW II that described the island-hopping campaign against Japan) including participants moving physical pieces on big boards (tabletop games), modern games employing computers and game theory, simulations that mimic the physics of modern weapons systems, and event-based simulations showing relationships between systems or parts of systems.

What is the difference between war at sea and any other SoS? Why use this as a lens? As mentioned previously, warfare in general and especially war at sea is a further manifestation of interactions between SoS and must also consider CASOF. The conditions of that theater of war are actually a co-evolution of the forces as they are employed strategically, operationally, and tactically. Adaptation to changing contexts in a theater of battle would have had to been part of the systems engineering prior to the beginning of conflict in which these systems are employed. Emergence comes out of the interactions between SoS, and their capabilities and limitations. And, what is the fight? One esteemed professor has called any coming war at sea as "a knife fight in a dark closet." Meaning it is likely to be fast and determinative, which makes the initial conditions even more important. During the rest of this chapter, we will bring in considerations (possibly behaviors) that may or may not be part of approaches to understanding emergence as a tool for designers and warfare planners.

4.3 KNOWLEDGE BEHAVIORS

Considering the complexity and possible positive and negative emergent behavior of a CASOF, knowledge is a necessary component for VASM Cebrowski's vision of shared awareness. When and how shared awareness is possible has its own set of interesting consequences. In its meaning here, knowledge is "actionable information."

The USN is quickly reaching the point at which its current, hardware-centric approach to warfare will become inadequate. Even today, legacy combat systems are unable to handle the torrential flows of information and knowledge required for combat success, and even worse, the current manner in which such systems are specified, designed, integrated, and tested fails to address the kinds of flexible, composable SoS that will be so critical for combat as future systems become operational. The research described in (Gallup, Nissen and Iatrou, 2019) examines how network-centric models and methods can be integrated with knowledge-centric models and methods to develop more appropriate approaches to combat system specification, design, integration, and testing. This helps to establish a powerful, unprecedented capability, the use and utility of which we illustrate through application to a challenging maritime scenario centered on carrier strike group operations at sea. Figure 4.1 below shows the dimensions of knowledge flows. Knowledge can be tacit (explaining how to ride a bicycle is difficult, but experiential knowledge makes it possible). Tacit knowledge flows slowly. The axis showing tacit to explicit knowledge acknowledges that explicit knowledge is different, in that it is written or otherwise passed (e.g., chat). The movement of tacit knowledge across the dimension of reach (how many individuals or organizations can use tacit or explicit knowledge as a function of time) is slow. Explicit knowledge passes quickly but is not as powerful as tacit, expert knowledge.

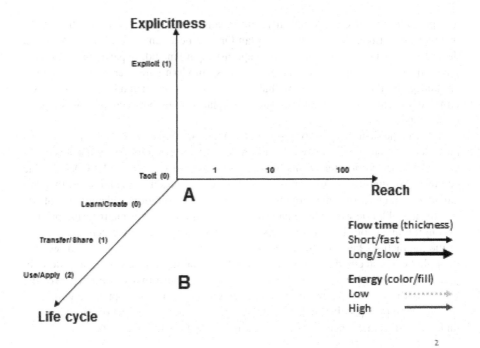

FIGURE 4.1   A three-dimensional view of explicit, tacit, life cycle, and reach dimensions of a knowledge system. (Nissen, 2018.)

The dimension of life cycle shows that knowledge is cyclic as it is learned, transferred, and applied. The A and B points in this diagram refer to the different ways in which knowledge can be transferred from A to B given the dynamics of the knowledge flow system.

What is the relationship to SoS and emergence? Assuming that we are approaching a capability of shared awareness across the battlespace, shared awareness itself requires some attention in systems design and inclusion among the factors being pushed into the various models (most importantly Monterey Phoenix (MP)) that predict emergence.

To demonstrate this, we have used basic physics equations as an analog to create a pseudo-mathematical view of a knowledge system, the table below shows a simplified version of the construct.

Recalling the basics of physics equations (Table 4.1):

Briefly, *knowledge force* (KF) is analogous to physical force and represents the effort required to accelerate knowledge in an organization. From knowledge flow theory (Nissen, 2018), it is expressed as a function of the knowledge *chunks* (C) being accelerated and the *explicitness* (E) of such knowledge (Table 4.2).

The lesson being expressed here is that knowledge, knowledge flow, knowledge power is an important part of the design phase of SoS. Once again, we cannot truly express the possible outcomes of emergent behaviors in a CASOF without

TABLE 4.1

Basic Physics Equations Used as Analogs for Knowledge Flow Equations

Construct	Description	Equation	
Force (F)	Effort required to accelerate Mass	$F = m \times a$	(1)
Work (W)	Force applied through distance	$W = F \times d$	(2)
Time (t)	Time for a mass to move its Distance	$t = \sqrt{(2d/a)}$	(3a)
Distance (d)	Distance that a mass moves	$d = \frac{1}{2}\,at^2$	(3b)
Acceleration (a)	Change in velocity	$a = 2d/t^2$	(3c)
Power (P)	Work done per unit time	$P = W/t$	(4)

TABLE 4.2

Knowledge Flow Equations (Nissen, 2018)

Construct	Equation	
K-Force basic	$KF = C \times (10 - 9E) \times o$	(5)
K-Force	$KF = (C \times pf \times KP_i) \times (int - \; sl \times E\textasciicircum nl)$	(5c)
K-Work basic	$KW = KF \times R \, (= KE)$	(6)
K-Energy	$KE = KF \times R$	(6a)
K-Work	$KW = KE \times E_i \times of$	(6d)
Flow Time	Measure	
K-Power basic	$KP = KW/FT$	(7)
K-Power output	$KP_o = (KW \times E_i \times KP_i)/FT$	(7b)
K-Power out extended equation	$KP_o = \left(Cpf\, of\, E^2\, KP^2\, R\right) \times \left(int - [sl \times E\textasciicircum nl]\right) / FT$	(7c)

acknowledging that shared awareness is an outcome of design, and that competitive advantage is gained when it is included.

4.3.1 Ignorance

Another knowledge behavior that affects SoS is the "taxonomy of ignorance" (Table 4.3).

> Ignorance can be construed as the state of there being corresponding knowledge 'out there' inaccessible to the decision maker. Holtzman (1989) identified several levels of ignorance by describing the equivalent knowledge paradigm within which it occurs. Table 1. presents an expanded version of the Taxonomy which identifies "Ignorance Level", "Description" and "Knowledge Required" separately for clarity. This should be useful as a starting point for mapping ignorance levels onto the decision-making situation.

(Denby and Gammack, 1999)

TABLE 4.3
A Taxonomy of Ignorance

Ignorance Level	Description	Knowledge Required
Combinatorial	Computational task too difficult, e.g., problem with 10^{40} variables	Mathematics model available; use of supercomputers
Watsonian	Cannot make the connection from all the clues; solution method incomplete	Method for determining the important facts from the unimportant ones, and drawing the right conclusion
Gordian	King Gordius tied a knot for the future king of Asia to untie. Alexander the Great was able to "untie it" by cutting the knot with his sword, thus solving the problem in an unusual way	Lateral thinking; are there "rules" to be broken?
Ptolemaic	Attributed to the Greek mathematician and astronomer, Ptolemy, whose model of the universe centered around a stationary earth	Evidence and observation of reality
Magical	"No one knows how it works, but everyone knows that it works," e.g., the use of Aspirin and other similar drugs	Trial and error
Dark	No model is available but one is aware of the issues, e.g., "What is life?," "Consciousness," etc.	Future of science
Fundamental	Unaware of issue (Ignorance is bliss!)	

Source: Denby and Gammack (1999).

In another article (Armour, 2000), five dimensions of ignorance are identified as

- "Lack of Ignorance" (lack of awareness of the state of ignorance)
- "Lack of Knowledge" (the state we are trying to fill)
- "Lack of Process" (a fear discussed by many that transparency is not possible in SoS)
- "Meta Ignorance" (similar to "not knowing what we don't know)
- "Synergistic Effects" (many ways to combine independent dimensions)

As we begin to note the different dimensions of SoS in CASOF, it becomes increasingly clear that this is an area of extreme dynamic volatility, in which information to knowledge sharing plays a crucial role.

4.4 STRATEGY, OPERATIONAL ART, AND TACTICS IN HYBRID WARFARE

The relationship between strategy and operational art links tactical action to strategic purpose. Operational art governs the deployment of forces and the arrangement of operations to achieve operational and strategic objectives.

Moving forward to the present and near future, the USN has become increasingly interested in autonomy and "intelligent" vessels of all kinds, i.e., air, surface, and subsurface. An autonomous vessel and its sister ship are now undergoing testing of their primary sensors to avoid other vessels in accordance with international rules of the road and potential tactics. Stating once again Maier's definition of emergent behavior as

> The system performs functions and carries out purposes that do not reside in any component system. These behaviors are emergent properties of the entire system of systems and cannot be localized to any components system. The principal purposes of the system-of-systems are fulfilled by these behaviors.

> *(Maier, 1998)*

Multiple perception systems (sensors) compare what is being sensed, use logical models, worldwide navigation maps, and rules of the road to provide inputs to the ship's navigation for the most acceptable maneuver. Some difficulty arises when a maneuver creates a need for consequent maneuvering. In other words, an immediate solution to a problem may induce additional problems to be solved downline. This amplification of effects is difficult to know ahead of time, especially in a volatile environment.

4.5   THE US NAVY AND EMERGING NEW AGE OF ENLIGHTENMENT

Here the discussion moves to the effects of environments and evolution of thinking in naval power. When human beings discovered they could move loads more easily using the innovation of a wheel, the next step was not a car. Radical ways to improve work or turn ideas on their heads (why didn't we think of that before?) or rethinking warfare doesn't happen in leaps. Opportunity, necessity, the arc of technical advances, economics, and the nature of threat (perceived or otherwise) are what shift the art of war and ways to conduct it.

Let's walk back the lowly wheel. From cart to chariot, to logistics support for armies, to moving artillery, to motor-driven personnel carriers and all sorts of vehicles. All dedicated to the fundamental need to move something to somewhere else.

No human in the stone age could have envisioned these things. Change happens by design and the factors previously mentioned.

We in the USN are in a position now to advance into a new age by design. Within the process of design are accumulations that are the independent pieces available now or in a near horizon to make things happen.

Enter the USN in near-peer competition to control important sea-lanes of communication (SLOCS), especially the island chains of the Indo-Pacific (and in the northern Atlantic). The treasure of our nation will only build, support, man, and maintain a certain number of currently commissioned and future ships, and a similar situation applies to aircraft, submarines, and other current weapons systems. This constraint creates a need for design options. One of these options is the surface autonomous vessel. Force structure dependencies are also complex. "Ideally, a country's naval force structure changes with national strategy, national treasure, technological advancement, and potential adversary capabilities. National strategy provides the rationale for, purpose of, and priority among choices to be made in creating a fleet. National treasure defines the resources and constraints dictating strategic choices. New technologies provide opportunities for increasing fleet effectiveness, yet also may endanger fleet survival should potential adversaries expose and exploit vulnerabilities in these technologies. This is a complex problem even when one takes into account only these four factors; however, USN acquisition also is challenged by other influences that inhibit capitalization of new technologies.

The most powerful of these inhibitions is inertia. The existing fleet represents a capital-heavy investment by the country, one with long build times and lifetimes. Ships and aircraft cost billions to design, build, and maintain. They require a capital-intensive industry featuring heavy equipment, infrastructure, and a skilled workforce—all generations in the making. As a consequence, annual programming and budgeting decisions are marginal in nature. It is the nature of a large fleet to evolve slowly, as opposed to undergoing revolutionary changes to its composition. This is a reality the Chief of Naval Operations (CNO) faces when considering changes to the naval forces. Each CNO's relatively short tenure restricts the ability to formulate, market, and execute any maritime strategy that would have a comprehensive effect on ship and aircraft procurement (Kline, 2017).

Unmanned surface vehicles and the science and engineering behind them are in their infancy. Much as the USS Langley and her experimental air wing evolved into the CVN and carrier air warfare of today, battlefield autonomy and artificial intelligence are new concepts and technologies that will need time and experience to evolve.

4.6   THE DYNAMICS OF DETERRENCE, THREAT, AND COMBAT

The following discussion is a brief overview of the idea of "control." What does it mean in a military sense? Sea power, taken as a whole, for example, is a "controller." Means of control are generally associated with mechanical devices, electronic feedback systems, and computer routines. The principles of cybernetics (control theory) are the same even when scaled to geopolitical control of SLOCS. Within this theory of control is the concept of variety of options. Controllers need to have as much variety as that which is being controlled. In routine systems, this variety can be established, and controllers work well within established limits. Think of a thermostat as a controller; it only needs three degrees of freedom to control the temperature in a room (on, off, and temperature setting). Deterrence as a form of a controller requires a variety of action, but the level of variety and need for control increases nearly exponentially with the potential threat of war, and nearly infinitely in combat. The problem is one of "emergent behaviors." These are the consequence of actions that are not expected nor within the range of control.

In the Medium Unmanned Vessel (MUSV), there are perception systems that compare what is being sensed, application of logic algorithms, and use international rules of the road to provide inputs to the ship's navigation to determine and perform the most acceptable maneuver.

We have not been in a combat situation with human-machine teams or autonomous vessels alone, but past experiences, e.g., the shootdown of an Iranian civilian aircraft (USS Vincennes incident) or the two 2017 collisions that left 17 sailors dead, that tell us something about possibilities of emergent behaviors at the low end of the technology, that is, where technology is an information enabler. Once again Ashby's Law of Requisite Variety (Umpleby, 2008) ("every good regulator of a system must be a model of that system") to assure, without fail, that emergent behavior is addressed if and when it is manifested. The "not knowing the future" in past events was ignorance coupled with "what might happen." In the coming age of autonomy, and in combat conditions, ignoring by human operators must go up (a consequence of too much information), along with trust in correct action by autonomy. But no one can be sure of the true consequences. For example, sparse control of unmanned vessels might break down. Humans, it turns out, are still the ultimate controllers in the face of the unknown.

The physical results for the presence of emergent behavior will be seen in the collective strategy of unmanned platforms working within the OODA loop associated with human planners. Additionally, it must be considered that SoS in military combat operations are interrelated by a complex web of possible outcomes that are increasingly unpredictable.

A few words here about the concept of SoS. Keeping in mind the definition presented at the beginning of this chapter, warfare engages in a higher level of SoS (CASOF) with theoretically greater span of consequences and emergent behaviors leading to those consequences.

Implications here could be devastating. Specifically, the synchronization of intelligence, weapons on target, and required effects could be entirely disrupted. Conversely, this emergent behavior could provide exactly what is needed in a very complex SoS during combat. Unfortunately, no one can be sure which way things will go.

Where emergent behavior will manifest itself is a bit more complex. The concept of the networked SoS described earlier may have new surprises for us. This new organization promises to allow the concentration of lethal power where and when it is needed. Where emergent behavior is most likely to show up is in the configuration of the combat network, targeting, and, eventually, weapons employment. How it will manifest itself will be in the lack of control by human supervisors with the implication that both good and bad effects are possible. A tricky question arises as effects of emergent behaviors are perceived by either the system or humans. First- and second-order effects were described by Ashby, as well as feed-back and feed-forward controls. First-order effects, those perturbations within an expected range of system behaviors, are easily corrected by controls within the system. Second-order effects are outside of the expected range and require additional capacity of the system to adjust back to a range of expectations. However, there is also a third order of effects, those that are so far outside of the range of expectations that the system is unable to use designed means to bring it back into stability. Control features include feed-forward and feed-back in systems. However, continual feed-forward will drive the system out of control, and feed-back can constrain the system to the point of not being effective. All of these concepts are interrelated within the Law of Requisite Variety. An example system that demonstrates these emergent behaviors is economics. Stimulus packages are feed-forward controls, and moving interest rates are feed-back. In general, the outcomes are known and expected, and can be tweaked to keep the system stable. Third-order effects would be the tumbling of the stock market. Still these emergent effects are well known. The Taxonomy of Ignorance (Denby and Gammack, 1999) states that the a priori knowledge of what might emerge and controls to maintain stability are separated into categories of what is known and not known, with the worst-case condition being the famous "we don't know what we don't know." This is the condition of combat, the extreme of requisite variety, and where decision of effects through feed-forward or feed-back comes down to intuition. Artificial intelligence (AI) is being honed to help with this condition, but it is not known if it can be of value in third-order emergent effects. An updated view of Cybernetics in this new age can be found in Umpleby et al. (2017)

> Whereas the social science disciplines create descriptions based on either ideas, groups, events or variables, cybernetics provides a multi-disciplinary theory of social change that uses all four types of descriptions. Cyberneticians use models with three structures – regulation, self-organization and reflexivity. These models can be used to describe any systemic problem. Furthermore, cybernetics adds a third approach to philosophy of science. In addition to a normative or a sociological approach to knowledge, cybernetics adds a biological approach. One implication of the biological approach is additional emphasis on ethics.

It is, however, possible that moving directly from manned to unmanned systems leaves holes in variety to meet the situational context and will not be adequate to account for third-order effects. Combat will always be more than a statistical game, but some principles have emerged from Admiral Horatio Nelson (command by negation and minimal instructions to commanders), Captain Alfred Thayer Mahan (overwhelming first strike and protection of sea lanes of communications (SLOCS)), and Captain Wayne Hughes (first and second strike wins in combat at sea). Combining the difficulties of uncertainty with the best possible way forward is the theme of this work. Commanders collaborate with their seniors and peers to resolve differences of interpretation of higher-level objectives and the ways and means to accomplish these objectives. Commanders generally expect their higher HQ has accurately described the operational environment, framed the problem, and devised a sound approach to achieve the best solution. Strategic guidance, however, can be vague, and the commander must interpret and clarify it for the staff. While national leaders and Combatant Commanders (CCDRs) may have a broader perspective of the problem, subordinate Joint Force Commanders (JFCs) and their component commanders often have a better perspective of the situation at the operational level. Both perspectives are essential to a sound solution. During a commander's decision cycle, subordinate commanders should aggressively share their perspectives with senior leaders to resolve issues at the earliest opportunity. An essential skill of a JFC is the ability to assign missions and tasks that integrate the components' capabilities consistent with the JFC's envisioned Concept of Operations (CONOPs). Each part of the art of Joint Command II-3 component's mission should complement the others (Joint Publication3-0 22 October 2018).

The commander's ability to think creatively enhances the ability to employ operational art to answer the following questions: (1) What are the objectives and desired military end state? (Ends), (2) What sequence of actions is most likely to achieve those objectives and military end state? (Ways), (3) What resources are required to accomplish that sequence of actions? (Means), and (4) What is the likely chance of failure or unacceptable results in performing that sequence of actions? (Risk). Operational art encompasses operational design—the conception and construction of the framework that underpins a joint operation or campaign plan and its subsequent execution. Together, operational art and operational design strengthen the relationship between strategic objectives and the tactics employed to achieve them (Joint Pub 3-0).

Many aspects of war at sea include their own set of system behaviors. For example:

- Technology emergent behaviors (platforms, weapons, C2, sensors)
- Leadership emergent behaviors (commander's intentions, or mission command)
- Emphasis on positive emergence (contributions of all toward success)
- Hazards of negative emergence (often the least understood of the behaviors

4.6.1 THE EFFECT OF THE "SINGULARITY"

The game of "Life" introduced by John Conway in 1970 (https://bitstorm.org/gameoflife/) creates apparent complexity out of just four rules. Populations emerge, grow, split, join, diminish, and die. The rules are the form of communication between the cells that bound their range of options and specify the outcome of a specific interaction. What creates the appearance of complexity is the interaction of multiple cells such that there is much greater possibility than that experienced by a single unit.

In the information revolution, we sometimes forget that our advance is based on only one simple construct. That is, a switch can be in two states, "on," and "off." It is the collection of the switches organized into units that creates the infinite possibility of information.

And, so it is with other self-organized systems, adapted to purpose that appears to be complex, but are ultimately reducible to a definable set of rules. Ant colonies, bee-hives, and other systems in nature are examples.

Does this work for human systems as well? What is different here? The system examples used so far are largely "autopoietic," or the result of autopoiesis a term coined by Humberto Maturana and Francisco Varela in "the Tree of Knowledge" (1987). That is, a system capable of replicating and maintaining itself. At the unit level, humans and other biological systems reflect this principle.

But, at the higher level of organization, what we have is human endeavor employing information in ways that a bee colony or ant hill cannot. That is, human systems can decide to adapt, not just in a response, but in ways that change the conditions for having the need to respond. The impact on environment is an example.

The concept of a singularity "The singularity is that point in time when all the advances in technology, particularly in artificial intelligence (AI), will lead to machines that are smarter than human beings." and contemplating a possible future before it has arrived is another. That is, humans creating the condition and then deciding what it means for possible future states that feed-back to the present with questions about what it means. So, self-reflection is possible in this system.

Getting to the question then, "in what ways should we (in particular the U.S. Navy) prepare for the coming singularity(ies), and what are the consequences for organizing in the present to bring forth that future?"

The singularity is not a line in the sand, a state that can be known as "pre-singularity" and "post singularity." It is a concept with a lot of people applying their own version of what it means. However, *something* is happening.

For one thing, the apparent rate of technical transformation seems to be increasing Kurzweil's Law of Accelerating Returns (Kurzweil, 1990), in particular areas and globally. Rather than following a roadmap for development however, the advances are emergent, and with that comes evolution in other areas, or at least the need to adapt.

The military is one place where adaptation and early adoption seem to occur. Seeking combat advantage over other potential adversaries is a never-ending system feed-forward and feed-back. It is here that we concern ourselves with what constitutes fundamentals so that we can re-use and advance concepts. Networking is one area, in particular, where technical advance has created a need for understanding the possibilities in design and applications.

The use of new and ever-increasingly capable autonomous systems is another example. The USN is currently experimenting with an autonomous vessel that can stay at sea for as long as 90 days and operate entirely independently and autonomously doing missions that are constrained only by policy restrictions and command and control questions (e.g., who "owns" the vessel).

The "Command and Control" (C2) question above is actually a fundamental. Robots, autonomy, and networks are employed within sets of rules and policies. They are currently separable. In the future (perhaps this is another notion of singularity), all of this will be blurred. In the future, the network is the platform, and robots/autonomous vehicles/manned vehicles will serve the network. In this future, what is organized, by whom, and for what purpose will be likewise very fluid. And the "whom" may not be a human, as the "hive" becomes increasingly complex. It will take computing power beyond human capability to plan and operate this network.

Is there an end-point to this future? In an article by Gaia (2014), a mutually dependent global society will continue to emerge with a shared set of beliefs "a civilization of ideas." It may be that we have the tools at hand to create this now, but between the present and the future, our task is to continue to understand the possibilities and re-think how we best use the emerging capabilities.

4.6.2 ADVANCEMENT OF AUTONOMY IN NAVAL OPERATIONS

In 2013 the DoD Unmanned Systems Integrated Roadmap 2013–2038 was signed out by then Vice Chairman of the CJCS, Admiral James Winnefeld and by Frank Kendall, the Undersecretary of Defense (AT&L). The document was recognition of the important role that unmanned systems will play throughout the DoD. A major contribution is the definition of key technology areas, and within each, objectives and activities relevant to continued development. Some specific technology areas and objectives and activities are important to the prototype autonomous ship, e.g., Interoperability: modularity to reduce costs; Communications: platform agnostic C4 systems, and plug and play capabilities; Security: protective measures to prevent compromise, remotely/autonomously render data at rest unrecoverable by adversary; Autonomy and Cognitive behavior: capability to perform dynamic tasks, not just pre-programmed ones, be able to modify strategies to meet human-defined goals.

Originally designated the ASW Continuous Trail Unmanned Vessel (ACTUV) but now called the MUSV moves the technology and operational vision of the integrated roadmap through the Defense Advanced Research Projects Administration (DARPA) defined ACTUV missions:

1. Explore the performance potential of a surface platform conceived from concept to field demonstration under the premise that a human is never intended to step aboard at any point in its operating cycle.
2. Advance unmanned maritime system autonomy to enable independently deploying systems capable of missions spanning thousands of kilometers of range and months of endurance under a sparse remote supervisory control model. This includes autonomous compliance with maritime laws and conventions for safe navigation, autonomous system management for operational reliability, and autonomous interactions with an intelligent adversary.
3. While the ACTUV program focused on demonstrating the ASW tracking capability in this configuration, the core platform and autonomy technologies were broadly extendable to underpin a wide range of missions and configurations for future unmanned naval vessels

Autonomy in naval systems is not entirely new. War at sea can happen fast, with a first strike being all important. For this reason, Phalanx guns and Aegis systems have autonomy built in. What is new is that the platform itself can now autonomously transport other capabilities that are themselves autonomous or are sparsely regulated by human intervention, employing networked information. This is the essence of a SoS design.

4.6.3 SEA CHANGE

War has become much less a mass of hurling mass and energy at the enemy than a matter of harnessing data-mostly in the form of guidance that has reduced miss rates and thereby the amount of mass and energy needed to hit a given target.

(Arquilla and Denning)

USS Monitor was necessary in its time to meet a specific threat—the ironclad CSS Virginia. Screw propulsion and steam power met with an advance in gunnery—the revolving turret, to make the USS Monitor capable of meeting the threat. While simple, and relatively unseaworthy, this vessel's advances made all other warships obsolete. It changed the way naval vessels would be designed and led the way to all other classes of naval vessels including the dreadnaughts, battleships, and modern warships of today. This single design had far-reaching effects, resonating through the history of naval warfare.

Today's naval forces are now engaged in meeting threats of swarming weapons, over the horizon missiles, and cyber-attacks on an information-infused battlespace, designed to deny access to satellite systems necessary to many warfare tasks. At the same time, advances in an emerging science of robotics and autonomy may serve as the revolving turret, changing everything yet again.

4.6.4 TECHNOLOGY AND INNOVATION

There is little doubt that we are in a period of rapid technology evolution. As an advance is made in one area, others begin to benefit and each advance becomes part of a fabric of interwoven and networked advances that seem to be accelerating the pace of change. Perhaps the reality is that networked connections between thriving technologies (e.g., the Internet of Things) are the true advance.

Technological innovation and military advancement are interconnected and create the synergy of rapid change.

"When the West began its ascent to world supremacy in the sixteenth century, military institutions played a crucial role in its drive to power. Recent historical work suggests that the Western military framework has undergone cyclical periods of innovation beginning in the early fourteenth century and continuing to the present and that such periods have resulted in systemic and massive changes to the basic nature of warfare and the organizations that fight. The military history of the twentieth century indicates that this pattern has continued unbroken except that the periods between major innovations have been decreasing even as the complexity of innovation has increased" (Murray and Millett 1996).

Then there is the oft-cited Moore's law where computing power doubles every 2 years, although Robert Keyes felt in 2006 that there are places where Moore's Law is "colliding with basic aspects of the Physical world." Denning and Lewis note that Moore's Law is really the confluence of chip (size and speed), system, and community. "Growth (progress) feeds on itself up to the inflection point. Diminishing returns then set in, signaling the need to jump to another technology, system design, or class of application or community" (Denning and Lewis 2017). Indeed, Intel seems poised to deliver a 10 nm chip sometime this year ("Intel Finds Moore's Law's Next Step at 10 Nanometers"). The larger point here is that what seems nearly impossible at one juncture is often overcome by shifts in technology within the will and purpose of a community.

> Today technology is an enabler of the revolution in military affairs, allowing changes that political and military leaders would like to make as they respond to political,

economic, and social changes. But it can also be an independent variable, forcing uncomfortable changes and, sometimes, eroding stability and order. New technologies or new combinations of technology have the potential to alter not only tactics and operational methods, but military strategy itself....Coming decades are likely to see the proliferation of robots around the world and in many walks of life. Hans Moravec, for instance, contends that mass-produced robots will appear in the next decade and slowly evolve into general-purpose machines. Ray Kurzweil takes the argument even further and holds that by the end of the 21st century, human beings will no longer be the most intelligent entities on the planet. However fast the evolution of robotics proceeds, it will invariably affect armed conflict. As one of the most avid customers of new technology, this will certainly affect the American military.

(Metz 2000)

4.6.5 A Sense of Urgency

Whether there is a perceived acceleration of change or a diversification of the portfolio of technology shared among many, the impact on warfighting is immense and should give rise to pressure, a sense of urgency where the need to capitalize on technology for strategic and tactical advantage can make all the difference.

If this is true, why does it take so long to manifest shifts in technology evolution? Even as advances are made, acquisition timelines are such that the time devoted to procuring revolutionary capabilities can be outpaced in the end by the development of new technology. What creates a sense of urgency? In his book, *A Sense of Urgency*, John Kotter (2008) explains that a true sense of urgency is rare, mainly because "it is not the natural state of affairs. It has to be created and recreated." So the task of leading a team of people in a transformation at any level will often require an ability to create an atmosphere of urgency that can be embraced and in turn bring about an atmosphere of achievement.

From the DoD Integrated Unmanned Roadmap (2013), there are pressures that create the need for autonomous systems:

1. Reduction in federal budgets;
2. Operational issues will be more complex;
3. US military forces will be re-balanced toward the Pacific;
4. Violent extremism will continue to threaten US interests;
5. Unmanned technologies will continue to improve; and,
6. Cyber domain will be a conflict environment.

In spite of these pressures, it is difficult to detect a sense of national urgency in developing autonomous capabilities in general or a rather large unmanned surface vessel, in particular. These capabilities are more seen as evolutionary rather than immediately revolutionary to the way we think about the strategic levels of war.

However, in the current geopolitical context, there are some factors that might be used to create the urgent need for technology adoption in the direction of unmanned surface vessels with autonomous capabilities:

First is the absolute need for more platforms, doing more of the essential missions being asked of the current naval force. Quoting from a USNI report:

> A historically small fleet and a relentless operational tempo are proving the Navy is too small to meet more than its bare minimum requirement around the world, Vice Chief of Naval Operations Adm. Bill Moran told a Senate panel on Wednesday (8 February 2017).

"We know we're too small for what we're being asked to do today," Moran told the Senate Armed Services subcommittee on readiness and management support.

"A smaller fleet operating at the same pace is wearing out faster. Work has increased, and we're asking an awful lot of our sailors and Navy civilians to fix [it]."

Second, urgency is created by the rapid development of autonomous systems in the US and other navies. The rate of developmental technology of autonomous systems is akin to the Moore's Law. For example, swarming capabilities have demonstrated 50, 100, 500 aerial vehicles operating together over the span of just a few years. Also is the sense of "If we don't do it, they will," and being left behind in the development and implementation of autonomous systems is inherently risky.

> Russia conducted a test of a revolutionary nuclear-capable drone submarine that poses a major strategic threat to U.S. ports and harbors. U.S. intelligence agencies estimate the Kanyon secret underwater drone will be equipped with megaton-class warheads—the largest nuclear weapons in existence, with the killing power of millions of tons of TNT. The weapon likely could be used against U.S. ports and bases, including those used by ballistic missile submarines.
>
> *(Gertz 2016)*

Third, it is the rise of the SoS that had its infancy in WWII. "Carrier aviation, strategic bombing, the integrated air-defense system that defeated the Luftwaffe in the Battle of Britain, and the amphibious warfare all appear to be instances of integrated, combined systems "military revolution exemplified by the Blitzkrieg of May 1940" (Murray and Millet 1996).

Fourth and finally, an obvious factor is that the network itself is advancing into a platform into which autonomy will plug into. Dr. John Arquilla and Dr. Peter Denning note that there is a recurring pattern in which

> Advances have been so swift and radical that just one or a few capital ships of the day could wipe out entire fleets of the previous generation's best warships....The rapid rise of digitization and networking signal the beginning of the end of the carrier's primacy. Indeed, by our reckoning, the next capital ship will be virtual. It will be a massive network of small digitally controlled entities, very artfully teamed with human operators.

Indeed, the network is the platform, and by extension is a warfighting capability, what Distinguished Professor Wayne Hughes calls "Information Combat."

4.6.6 CULTURE AND POLITICS OF ARMED CONFLICT

An entire treatise on this topic is beyond the scope of this report, but there are some very important points to be made relevant to forward force structure and therefore

SoS and CASOF. Doctrinally, the Chinese employ a centralized approach to warfare. Large amounts of information are processed, something we have experienced and are working through. However, the Chinese military doctrine displays a strong belief that strategy is a science rather than an art. It is important that the US continues to improve information dominance and learn to use it effectively and quickly by human decision-makers that can then write intentions that move the Chinese leadership into a lack of confidence in the outcome of hostilities. This is the core of deterrence.

The Chinese order of battle includes strike and bomber aircraft, ship-launched J-16 strike fighters, ground-launched short-range ballistic missiles (first island chain defense), and air- and ground-launched anti-ship cruise missiles (long range into second island chain). In 2017, naval forces consisted of 28 destroyers, 47 frigates, and 41 submarines able to launch Anti-Ship Cruise Missiles (ASCMs). Since then, they have significantly added to their inventory with a new aircraft carrier, additional amphibious ships, and much more. China also has a large inventory of mines of different types, adding blockade to their capability.

4.6.7 BASIC CONCEPT OF OPERATIONS (CONOPS)

The autonomous ship Sea Hunter was (and still is) a prototype vessel that is an engineering marvel, a breakthrough technology, and worthy of inclusion in our future view of war at sea. DARPA was simply provided some reasonable requirements on which to develop a prototypical design. Autonomy demonstrating navigation and adherence to rules of the road was the goal. In reality, after the vessel was built, advanced through sea trials, and performed some demonstrations under the Office of Naval Research (ONR), it was delivered to a newly commissioned organization for the development of autonomous ships, the Surface Development Squadron-1 (SURFDEVRON ONE) via Naval Information Warfare Command, Pacific (NIWC PAC) with an expectation that it was ready to be introduced to the Fleet.

Its introduction is ongoing, and there is now a CONOPS under development that answers some of the mission employment questions. While advancing in its mission set, its demonstrations have been as an additional arm to a carrier battle group and thus an expectation that it will be supervised by a larger vessel to ensure it manages tactics that have yet to be entirely worked out. In its fully developed form, with advanced sensors and long range, it may be freer to roam, performing isolated tasks, but those are largely in the future. And, at present, the Navy is very leery of arming a truly autonomous vessel. This is sensible, given the variety and emergence problem discussed at the beginning of this work and other problems such as cyberattack and security of an unmanned vessel.

4.7 LIGHTLY MANNED AUTONOMOUS COMBAT CAPABILITY (LMACC)

The LMACC concept is a middle mark in the evolution of autonomy in warfare. To state again, briefly: it is a high-speed vessel carrying long-range surface-to-surface missiles and anti-air missiles as its primary weapons, intended to be a presence in the first island chain in peace, or, in combat, using EW and tactics described below to become a very difficult target to find and hit. During the opening phases of war, its mission is first strike if needed or counterstrike against land missile batteries that threaten our heavy formations with their aircraft and large crews. This will create space and time for these assets to come within striking range of their targets to deliver decisive blows and shape the course of the conflict. As we move into the steady-state attrition and blockade, it will serve as part of an elastic anti-ship defensive line to prevent breakouts, transport Marines and supplies to control key terrain, and board commercial ships to enforce our blockade. Finally, since the Navy will always have non-combat obligations, a secondary objective of the LMACC program is to provide large numbers of affordable platforms for peacetime presence, patrol, and diplomatic functions.

As the USN moves into the unmanned age and implements Distributed Maritime Operations, there is a need for small, lightly manned warships to streamline that transition and fill roles that require a human crew. Congress has expressed concerns about unmanned vessels on a number of fronts and highlighted the need for a class of ships to bridge the gap. The Naval Postgraduate School's LMACC program has designed a warship to meet this need.

The need for these small, heavily armed warships has also been well established and is based on extensive analysis and wargaming across the Navy's innovation centers. These ships will provide distributed forward forces capable of conducting surface warfare and striking missile sites from within the weapons engagement zone of a hostile A2/AD system. They will be commanded by human tactical experts and operate in packs with supporting unmanned vessels, like the *Sea Hunter* MUSV, to distribute capabilities and minimize the impact of combat losses. But to show the possible positive and negative emergence, a scenario must be employed that tests the relationships between SoS as the context, intentions, and results become immediate feedback to the larger "regulator."

We begin with a story to derive the first iteration of a model. The great power competition (US, China, and Russia) has increased tensions as the world has moved into economic zones of influence. This includes our zone of interest here, the western pacific and China's continued power projection into international waters that it now claims as part of its territory. This is due in part to the development of man-made islands, dredged from the sea bottom and large enough to create a runway and add missiles and ports for ships. In addition, through still an island nation, Taiwan is now counted by China as part of its territory, adding another dimension to already very tense relations in the region. Along with economic and territorial claims, these actions contain the SLOCS, through which a very large proportion of world trade navigates on their way north through the Philippine Sea north and east toward the United States. Control of this SLOC gives the Chinese government a fait accompli, meaning that the free flow of resources can be cut off at any time, and the USN and American allies in the region can do little to prevent it.

In this context the USN is vulnerable, having very large logistics vessels and warships, with logistics requirements to sustain them at sea. They are also very visible and subject to effective attack by land- and sea-based long-range surface-to-surface missiles. This creates the need for a new strategy, one in which a persistent force of small warships that are lightly crewed (LMACC) and heavily armed with long-range missiles and shorter-range missiles for engaging ships are paired with unmanned vessels (MUSV) to create a networked sensor and weapons capability. The MUSVs are capable of autonomy, maintaining rules of the road, and can also be controlled or instructed to different positions and missions from the LMACC. Through this distribution of capabilities, the problem now becomes hard for the Chinese, in that it will be forced to waste their limited missile capabilities against unmanned targets and against heavily defended LMACCs.

Into this context, we begin to weave the SoS that are the backbone of navy information structures and planning processes. An intelligence network, strategic intentions, required effects (not everything is blown up, but perhaps interfered with in some useful way), and orders relayed to this LMACC flotilla (five LMACCs, five MUSVs), either directly or through a carrier battlegroup.

This is where we turn to MP and its ability to trace all of the possible paths through SoS interactions. We find that there are many more possibilities. Some will be unworkable, some that are unremarkable, and some that will give pause to the human reviewing possibilities.

But first, we must build a "base model" of the SoS interactions. The diagram of interactions is shown in block form below. It is the "story" of what is supposed to be interconnections of the extremely complex military system that is the USN (we limit this to just the USN's surface fleet on purpose as including all dimensions of air, sea, undersea, and land would take several chapters to explain but will perhaps be necessary to describe the entire SoS eventually). If we expand each of the blocks, we will see technical, organizational, intelligence, and specific tools employed for their work, as well as communications connections to other blocks. We will stay at the higher level as each block will represent all these SoS capabilities, except where we wish to vary some specific aspect of change.

4.8   GENERAL DESCRIPTION OF A SIMPLE MP MODEL

Two vessels USN vessels are involved. One is the LMACC lightly crewed, discernment, perception of environment, understanding commander's intent, and able

Laydown

- A lightly crewed warship is in company with an autonomous vessel.
- Conducting freedom of navigation operations (FON OPS).
- Intentions from higher authority are clear. Operate within the assigned Sea Lane of Communications (SLOC) to permit freedom of passage,
- DOMA is a big country with a long coastline. The country of ABBYSA is 150 miles off of their coast. DOMA lays claim to waters that include ABBYSA, which is separate country with its own claims. DOMA also claims that ABYSSA is part of their country.
- DDGs have been doing this mission in the past. Now we will try the LMACC plus an unmanned vessel to do this mission.

FIGURE 4.2   Geographic and political description of contested waters.

to fix malfunctions in the autonomy that guides the LMACC. Here is where the human-machine partnership is most important.

The second vessel is the fully autonomous MUSV (no crew), which can move over the horizon with a variety of sensors, sending sensor data back to the LMACC.

We are in phase 0 (peacetime operations) but engaged in near-peer competition with China which has capabilities and demands that include international waters. One navy mission is to do Freedom of Navigation Operations (FON OPS). Usually this is accomplished with a very expensive guided-missile destroyer (DDG) with

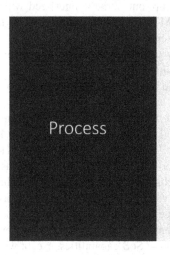

FIGURE 4.3   Description of process.

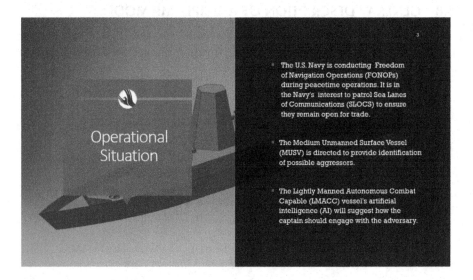

FIGURE 4.4   Initial conditions in operational situation.

Typical Event Flow Narrative & Model
High Level

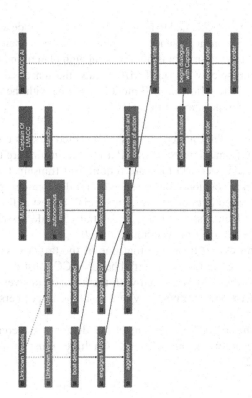

1. The MUSV (Medium Unmanned Surface Vessel) executes an autonomous mission.

2. The MUSV detects target vessels.

3. The Unknown Vessels can take various actions. In this case they aggressively engage the MUSV.

4. The MUSV sends the gathered intel to the Captain of the LMACC (Lightly Manned Autonomous Combat Capability) and the LMACC AI.

5. The Captain of the LMACC and its onboard AI receives the intel and actions of the unknown vessels from the MUSV.

6. The LMACC AI begins dialogue with the Captain.

7. The Captain partakes in dialogue with the LMACC AI.

8. The Captain issues or approves the order.

9. The order is sent to the MUSV and the LMACC AI and when the order is received, it is executed.

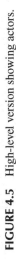

FIGURE 4.5 High-level version showing actors.

little autonomy. The SLOCS need to be constantly open for trade and free navigation, which is why the USN has a presence—keeping the SLOCS open.

What we are interested in here is not the campaign plan of the FON OPS, but the interaction between the crew vessel (LMACC) and the sensor vessel under possible varying conditions. The following MP model was done with the assistance of NSA intern partners during the summer of 2021.

> **Conditions:** Initial "normal conditions" All sensed vessels are showing AIS (automated information system, similar to how aircraft are tracked by controllers. Vessels over 100 tons are required to transmit this information, which includes position, course, speed, and destination), possible hostile vessels are sensed but not a threat. Of special interest to the LMACC crew is the idea of being "spoofed" by a potentially hostile vessel managing the AIS data differently or not providing it at all.
>
> **Change in Context:** It turns out that one of the target vessels is spoofing AIS—using a code to hide itself. The LMACC must decide on what to do with the other LMACCs and with the autonomous vessel (MUSV).
>
> **Use of Enhanced Autonomy:** Crew and machine in partnership agree on a course of action.
>
> **Emergent Behavior:** Crew and machine disagree on course of action. A dialogue between human and machine must resolve the issue (Figures 4.2–4.9).

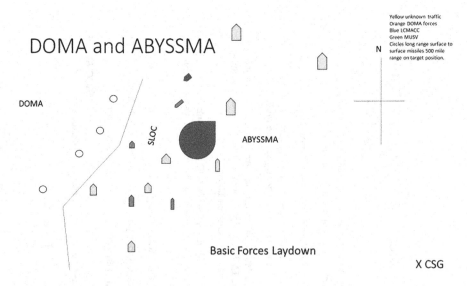

FIGURE 4.6   Basic situational configuration.

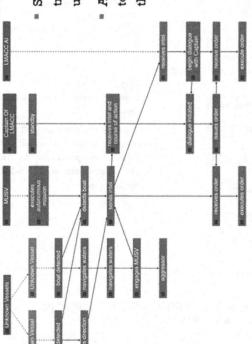

Alternative Event Flows and Results – High Level

- Scope #2 trace #35 in travel_contested_waters_high_level.mp, created using MP-Firebird

- Assuming that both unknown vessels are working together, one vessel retreats and the other one stays the course to engage the MUSV.

FIGURE 4.7   Example of emergent behavior.

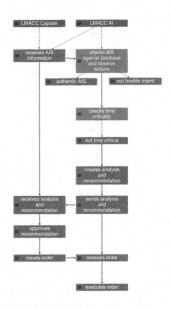

Typical Event Flow Narrative & Model Decision Making

1. The LMACC Captain receives AIS information from the MUSV (as shown in the previous model).

2. The LMACC AI checks the AIS against a database and observes the actions of the detected vessel.

3. The LMACC AI confirms the AIS is authentic and does not observe hostile intent from the detected vessel.

4. The LMACC AI then checks the time criticality of the situation.

5. The LMACC AI determines that the situation is not time critical.

6. Based on the findings, the LMACC creates an analysis and recommendation and sends that to the LMACC Captain.

7. The LMACC Captain receives the analysis and recommendation from the LMACC AI.

8. The LMACC Captain approves the recommendation and issues an order.

9. The LMACC AI receives the order and executes it accordingly.

*Note: Depicted is Scope 1 trace 3 in travel_contested_waters_decision_makingv2.mp

FIGURE 4.8 Typical decision-making including human-machine dialogue.

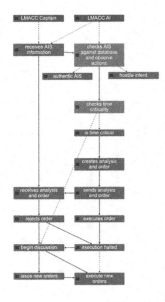

Alternative Event Flows and Results - Decision Making

- Scope 1 trace 8 in travel_contested_waters_decision_makingv2.mp, created using MP-Firebird

- The trace shows the LMACC Captain rejecting the order the LMACC AI created, halting its execution. Then, the two parties begin a discussion on what the new order should be with the LMACC Captain having the final say.

FIGURE 4.9 Emergent behavior as LMACC captain rejects the analysis of the machine partner and issues new orders.

4.9 SUMMARY AND COMMENTS

We have covered a lot of territory in this chapter. Questions about the definition and roles of autonomous systems have become increasingly important as the science and engineering of machines that can team with human counterparts has evolved. The complexity of SoS is not necessarily the generative principle that produces emergence. In our simple example, we found emergence arising from a few interactions. Yet, it seems something more is needed. The various themes included in this chapter point to ways in which SoS in war at sea (Complex Adaptive Systems of Opposing Forces) to include various configurations of autonomy might only really be understood as a human-machine partnership. Though not dealt with in depth here, the role of a dialogue between human and machine and an adaptive capability that looks much like the human autonomic nervous system might show us a better way to creating the lowering of complexity and the increase in variety needed to win. What is proposed here is the concept of human-machine *partnership* inclusive of teaming but with increased capability needs to be further defined and new means to move in this direction requires a working concept with which to guide the discussion. This is a research direction now being undertaken by this author with the help of a PhD student.

REFERENCES

Armour, Phillip G. "The Five Orders of Ignorance; Viewing Software Development as Knowledge Acquisition and Ignorance Reduction." *Communications of the ACM*, 2000, Vol. 43, No. 10, pp. 17–20.

Denby, Ema, and John Gammack. "Modelling Ignorance Levels in Knowledge-Based Decision Support." In *Proceedings of 2nd Western Australian Workshop on Information Systems Research*, 1999.

Denning, Peter J., and Lewis, Ted G. "Exponential Laws in Computing Growth." *Communications of the ACM*, 2017, Vol. 60, No. 1, pp. 54–65.

Gaia, Vince. Homni: "The New Superorganism Taking Over the Earth," 2014. http://www.bbc.com/future/story/20140701-the-superorganism-engulfing-earth?ocid=ww.social.link.email.

Gallup, Nissen, M. E., and Latrou, S. J. A knowledge based analysis of information technologies required to support fleet tactical grid. Monterey, California. Naval Postgraduate School, 2019.

Gertz, Bill. "Russia Tests Nuclear-Drone Sub." The Washington Free Beacon. Dec 8 2016.

Giammarco, Kristin. "Practical Modeling Concepts for Engineering Emergence in Systems of Systems." In *System of Systems Engineering Conference (SoSE), 2017 12th*, pp. 1–6. IEEE, 2017. http://hdl.handle.net/10945/61679.

Holtzman, Samuel. *Intelligent Decision Systems*, 1989.

Hughes, Wayne P., Jr. "A Salvo Model of Warships in Missile Combat Used to Evaluate Their Staying Power." *Naval Research Logistics (NRL)*, 1995, Vol. 42, No. 2, pp. 267–289.

Keyes, Robert W., IBM Watson Research Center, Yorktown, N.Y. reprinted in IEEE Newsletter, September, 2006.

Kotter, John P. "A Sense of Urgency." *Journal of Technology Case and Application Research*, 2008, Vol 10, No. 3, pp. 93–96.

Kline, Jeffrey E. "Impacts of the Robotics Age on Naval Force Design, Effectiveness and Acquisition." *Naval War College Review*, 2017, Vol. 70, No. 3, pp. 63–78.

Kurzweil, Ray. *The Age of Intelligent Machines*. Cambridge, MA: MIT Press, 1990.

Lanchester equations: https://en.wikipedia.org/wiki/Lanchester%27s_laws

Maier, Mark W. "Architecting Principles for Systems of Systems." *Systems Engineering*, 1998, Vol. 1, No. 4, pp. 267–284.

Maturana, Humberto R. and Varela, Francisco J. *The Tree of Knowledge: The Biological Roots of Human Understanding*. New Science Library/Shambhala Publications, 1987.

Metz, Steven. *Armed Conflict in the 21st Century: The Information Revolution and Post-Modern Warfare*. Army War College (U.S.). Strategic Studies Institute: Carlisle, PA, 2000.

Musashi, Miyamoto. *The Book of Five Rings*. Translated by Yagu Munenori. 1st Weatherhill ed., Boston, MA, 2006.

Murray, Williamson and Millet, Allen R. *Military Innovation in the Interwar Period*. Cambridge, UK: Cambridge University Press, 1996.

Nissen, Mark E. "Initiating a System for Visualizing and Measuring Dynamic Knowledge." 2018. http://hdl.handle.net/10945/58375. The article of record as published may be located at http://dx.doi.org/10.1016/j.techfore.2018.04.008.

Umpleby, Stuart A. "Ross Ashby's General Theory of Adaptive Systems." Department of Management, The George Washington University, 2008.

Umpleby, Stuart, Wu, Xiao-hui and Hughes, Elise. "Advances in Cybernetics Provides a Foundation for the Future." *International Journal of Systems and Society*, 2017, Vol. 4, No. 1, pp. 29–36.

5 Applications of Defense-in-Depth and Zero-Trust Cryptographic Products in Emergent Cybersecurity Environments

Kent D. Lambert
BlockFrame, Inc.

CONTENTS

5.1 INTRODUCTION

Cyberattacks on computer networks and cyber-physical systems have emerged as primary national and international security concerns. According to the National Cyber Security Centre of the United Kingdom, the loss of sensitive, private data, and revenue by governments, businesses, and individuals take place through a wide variety of disruptions from ransom payments, loss of access to data, and damage to computers and software (National Cyber Security Centre 2019). Organized crime and state-supported actors are the leading offenders, and public-sector targets are the victims of more than half of recorded incidents (Verizon 2019). According to the United States Cyberspace Solarium Commission, "The reality is that we are

DOI: 10.1201/9781003160816-7

dangerously insecure in cyber. Your entire life—your paycheck, your health care, your electricity—increasingly relies on networks of digital devices that store, process, and analyze data. These networks are vulnerable, if not already compromised. Hundreds of billions of dollars have been lost to state-sponsored intellectual property theft and cyber espionage. A major cyberattack on the nation's critical infrastructure and economic system would create chaos and lasting damage exceeding that wreaked by fires in California, floods in the Midwest, and hurricanes in the Southeast" ("Cyberspace Solarium Commission," n.d.). In early 2018, the United States Federal Bureau of Investigation and the National Security Agency released a rare joint report directly accusing the Russian government of active cyberattacks targeting components of the United States energy and other civilian critical infrastructure (FBI/NSA 2018). The sobering reality of the magnitude of the cybersecurity threat to United States federal networks, critical infrastructure (D. Trump 2017), and bulk power distribution systems (Donald Trump 2020) has prompted active research and development to counter strategic threats from cyberattacks.

Beyond the threats to business and government infrastructure, mass breaches of personal and financial data from businesses and governments have become a regular occurrence. In 2015, the United States Office of Personnel Management (OPM) reported a breach that resulted in the loss of personal information from over 21 million government employees (Office of Personnel Management Security Resource Center, n.d.). In 2018, the General Services Administration (GSA) announced that they had discovered a breach of the System for Award Management (SAM) portal, which affected 70,000 government contractors. In 2019, breaches of the Facebook data system may have compromised the personal data of 540 million users (Utermohlen 2019). The complexity of the system of systems (SoS) architectures of modern data enterprises contributes to their own vulnerabilities.

5.2 DEFINING CYBERSECURITY THREATS AS EMERGENT BEHAVIORS

Rainey and Tolk (2015) provide definitions for four levels of Emergence Complexity in complex systems: Simple Emergence, Weak Emergence, Strong Emergence, and Spooky Emergence. These definitions are useful to help characterize the high degree of Emergence Complexity from cybersecurity threats. Figure 5.1 (Mittal and Rainey 2015), graphically displays those concepts as an Emergence Complexity Cone. Increasing complexity is shown on the y-axis; the Cone is divided into deterministic and stochastic search-space domains on the x-axis; and the Cone volume depicts variety. The Cone perimeter forms a Constraint Boundary, while the Boundary of Knowledge is shown as a cylinder that exists between the field or variety between the constraints. The Boundary of Knowledge Cylinder around simple and weak emergence in the deterministic domain signifies that ample knowledge is available for subject matter experts (SME) to develop abstractions using known techniques like cybernetics, systems theory, control theory, or network theory.

The rapid and seemingly unchecked expansions of cybersecurity threats indicate that they exhibit behavior consistent with Strong or Spooky Emergence, with

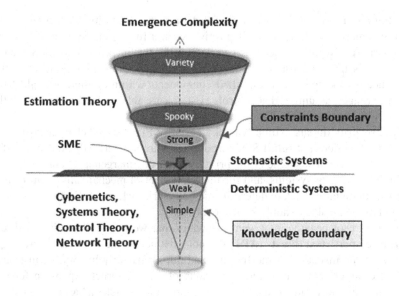

FIGURE 5.1   Mittal and Rainey emergence complexity cone.

characteristics of unpredictable or stochastic complexity. The general failure to successfully predict and contain the emergence of cybersecurity threats to a deterministic domain using the best-known technologies, models, and control theories helps validate that categorization. The presence of unpredictability and uncontrollability is also consistent with Mittal and Rainey's observation that, "when the knowledge itself does not exist to identify the variety and constraints [of the system], the behavior is classified spooky" (Mittal and Rainey 2015). Borky and Bradley make very similar observations about complexity within a systems engineering context: "We use the terms, 'complex' and 'complexity' in the sense of describing something that cannot be adequately dealt with using only simple, intuitive tools and methods"; and "Complex systems routinely exhibit behaviors that are difficult to predict or even to understand" (Borky and Bradley 2019). What is at issue for Emergent Complex system behaviors is whether or not they can be understood and/or controlled. If systems exceed the Boundaries of Knowledge of SME, those system behaviors are axiomatically less understandable and predictable. As system behaviors exceed the Boundaries of Knowledge and approach the Constraint Boundaries, they should demonstrate higher levels of Emergence Complexity, become more stochastic, and therefore less controllable. Just as the levels of Emergence Complexity can provide at least a qualitative description of the growth of those behaviors, they should also provide a qualitative framework for judging whether those levels of Emergence Complexity are increasing or are being reduced. This chapter seeks to transfer these concepts to cybersecurity threats. By classifying a cybersecurity threat as a level of emergence, we can better understand the characteristics of cybersecure systems and better design and measure their controls, constraints, and environments to control emergent threats and improve cybersecurity performance by mitigating threat effects.

Mittal and Rainey also provide useful definitions of "positive," "negative," and "neutral" emergence: Emergence is positive when it fulfills the System of Systems' (SoS) purposes. Negative emergence does not fulfill the purposes of the SoS. Neutral emergence is neither positive nor negative. The purpose of attempting to control emergence is to keep systems and their components within optimal or healthy performance ranges, assuming that controls exist that are able to affect emergence either positively or negatively.

Cybersecurity threats exhibit "Spooky Negative Emergence" characteristics, i.e., emergence that does not fulfill SoS purposes, and are not currently subject to effective control mechanisms. One of the primary goals of Emergence Theory is to identify highly emergent systems, and then to apply control mechanisms to move from stochastic boundaries back to deterministic boundaries where SMEs are knowledgeable and might be able to apply known controls.

The nature of negative emergent behaviors that we now call malicious threats or advanced persistent threats (APT) did not exist before the Internet. Therefore, at the time of the invention of the Internet and during its early history, security controls were largely not considered. The value of rapid Internet expansion for economic growth and rapid inexpensive free expression is undeniable, but it also left many users reliant on legacy systems with huge vulnerabilities. According to the President's Commission on Enhancing National Cybersecurity:

> No one—not even the visionaries who created the Internet a half century ago—could have imagined the extent to which digital connectivity would spur innovation, increase economic prosperity, and empower populations across the globe. Indeed, the Internet's origins in the defense community are today almost an afterthought, as its explosive growth has given it a dramatically different shape. Its creators could not have dreamed of the way and the extent to which our national and global economies have thrived, how innovations have been enabled, and how our population has been empowered by our digital connectivity. With these benefits and transformational changes have come costs and challenges. The interconnectedness and openness that the Internet, digital networks, and devices allow have also made securing our cyber landscape a task of unparalleled difficulty.

Commission on Enhancing National Cybersecurity (2016)

Worldwide Internet growth without planned or embedded cybersecurity might have been the first cause for the negative emergent behaviors because it was open to malicious exploitation. Additional negative emergent behaviors then grew from the growth of the threats themselves. Layered Defense-in-Depth (DiD) and Zero Trust Architectures (ZTA) are concerted reactions to cybersecurity threats, have certainly helped, and have not been totally successful in enforcing cybersecurity or trust, especially in legacy systems that were not originally designed for security. Ideally, all new systems should be designed with integrated cybersecurity components and processes, but technologies are also needed with "backward compatibility" to protect legacy systems until they can be replaced. The Eco-Secure Provisioning (ESP)™ Framework and the Philos™ blockchain can help reverse those negative emergent behaviors, and MBSE and MBSAP should be able to design, model, simulate, and measure their positive effects.

5.3 CONTROLLING NEGATIVE EMERGENCE OF MALICIOUS CYBER ACTIVITY

The negative effects of an initial lack of designed-in cybersecurity, compounded by emergent cybersecurity threats, have prompted transnational and cross-industrial responses. The recommended strategies to counter emergent cybersecurity threats are layered DiD, ZTA, and secure cryptographic key provisioning. DiD is defined as "deployment of diverse and mutually supportive security controls at various points in the architecture to create concentric shells of protection that an attacker must defeat to reach protected assets" (Borky and Bradley 2019). ZTA may be defined as "a strategy that extends and refines DiD architectures with fine-grained segmentation, strict access control, strict privilege management, and advanced protective devices, all based on a concept that no person and no resource is trusted without adequate verification." Within DiD and ZTAs, the BlockFrame (ESP)™ Framework and the Philos™ blockchain can provide several other techniques to enhance DiD and Zero Trust, including the Proof-of-Origin (POO) for data elements, uniformity of network segments in the event of loss of wide-area communications, multifactor identification (MFI) of components within Zero-Trust Networks to assure all remote components are verified, strengthening Transport Security Layers (TLS) by assuring the trust of certificates, and management and verification of software patches and version updates. ESP™ can also provide secure cryptographic key provisioning for even higher degrees of trust and security, particularly for distributed Internet-of-Things (IoT) devices.

DiD was a relatively simple, widely practiced, and initially successful architecture strategy to develop and secure cyber systems. In practice, quantitative studies, like the Verizon, "2019 Data Breach Investigations Report," now show that as a control mechanism for the emergent growth of cybersecurity threats and their risks, DiD alone has proved inadequate to secure cyber systems across government, industrial and private sectors (Verizon 2019). John Kindervag, Principle Analyst at Forrester Research, Inc., created the ZTA (Pratt 2018) in 2010, which was subsequently commercialized by Forrester. Zero Trust recognized that DiD alone did not adequately stop the exploitation of systems that had already been penetrated, insider threats, threats from social engineering, and other situations where trust might have been assumed, but was demonstrated not to be trustworthy. ZTA emphasizes the importance of partitioning the interior of vulnerable systems, the authentication and authorization of personnel, and other security practices under the assumptions that no personnel, hardware, software, or practices should be trusted until it is confirmed and verified as trustworthy (Palo Alto Networks n.d.). According to Borky (Borky and Herber 2020), DiD is still necessary but is no longer sufficient against APT. That is because cyberattackers are either already inside most systems, possibly resourced by hostile national governments, or because a ZT strategy would assume that they are. When ZTA complement DiD with state-of-the-art products within robust and resilient cyber designs, they represent the best solutions available today to APTs.

Although the combination of DiD and ZTAs are generally considered to be the best available cybersecurity design strategies for systems architectures, in practice the complexity of the cybersecurity threat environment, the complexity of the defensive DiD and ZTA architectures, and the known continuing damage from penetrations, data theft, ransomware attacks, and denial-of-service attacks continue to exist as negatively emergent phenomena. If Zero Trust assumes that systems have already been compromised, then one essential requirement for ZTAs should be the capability to reset or recover from a breach or compromise and to return to a trusted state. The huge proliferation of available security products still might conflict with one another, might not provide total security for all the gaps among the different products, or might not take a systematic and uniform approach to identifying, correcting, and monitoring trust levels throughout networks or enterprises (Gorog and Lambert 2020). Those "unknown" gaps, overlaps, and vulnerabilities still exist and demonstrate that current defenses cannot be trusted without more support from automated enforcement and governance, and better methods to reset or recover to a state of trust following compromises. If any ZTA cannot reset or recover, that should be identified as a systemic weakness that should be mitigated. New technical methods are being developed to help mitigate those trust problems, to enforce trust, and to recover trusted-state conditions from any compromise. Those approaches include secure cryptological key provisioning, blockchain-assisted logistics management, and the use of the improved architecting and design frameworks implicitly available through Model-Based Systems Engineering (MBSE) and supported by the Model-Based Systems Architecture Process (MBSAP). Agile Systems Engineering (ASE) architectural approaches, automated development and testing through Continuous Integration/Continuous Delivery (CI/CD), Test-Driven Development (TDD), and integration with rapid and scalable trust chain support are facilitating those developments.

A new set of applications are currently being developed and tested by BlockFrame, Inc., including the "Eco-Secure Provisioning™" (ESP™) supply chain security enterprise, secure cryptological key provisioning, and the Philos Blockchain/Distributed Ledger™ Enterprise (Gorog 2019). The unique BlockFrame® inventions and intellectual property explained in this chapter are protected by pending United States and international patents ("Patents Filed by BlockFrame, Inc. U.S. Patent Application No. 17/488,529; No. 17/488,562; No. 17/488,589; No. 17/488,655; Patent Cooperative Treaty (PCT) Application No. PCT/US21/52839; No. PCT/US21/52854; No. PCT/US21/52865; No. PCT/US21/52872." 2021). The ESP™ logistics security enterprise will use a cloud-based data architecture to connect participating objects through embedded, encapsulated, loosely coupled, and modular components. ESP™ secure cryptographic key provisioning has been successfully implemented and demonstrated for the United States Navy (Gorog and Lambert 2020). Philos™, a community-based open architecture blockchain, was initially planned to meet needs defined in Colorado Senate Bill SB18–086 (Lambert et al. 2018). The Philos™ distributed ledger system is designed as a cryptology-intensive, globally scalable enterprise, based on a series of encapsulated common units to connect components across its architecture. The common units have specialized cryptographic generation and storage nodes, secure data, cypher delivery, and a global neutral governance structure, similar to the Domain Name System (DNS) of the Internet ("Domain Name System" n.d.). The successful application of blockchain logistics management using the ESP™ and Philos™ will increase trust and lower risk among all participating users and will include robust and verifiable cyber-security designs for data confidentiality, integrity, and availability (CIA), zero-trust, data immutability, and data non-reputability.

According to Gorog and Boult (2019), "In a distributed ecosystem of devices, complete positive identification would require having undeniable POO on each data item from every device." ESP™ secure cryptographic key provisioning and Philos™ are all specifically designed to provide both hardware and data MFI and POO. Those built-in capabilities will also provide for rapid resets and recoveries for system faults and data losses back to a state of trust, and the capabilities to prove the identity of objects within the IoT through MFI, to secure data elements within supply chains, to secure public documents, to secure power grid data and operations, and to secure personal privacy information (PPI). They are also designed as open systems architectures, for the secure and effective delivery of secret cypher codes, and sound cryptographic practices. In particular, extensive externally managed ("out-of-band") MFI of actors, hardware, software, or almost anything of unproven trust is a key capability for successful integration and uniformity. MFI of system components, authentication of personal identities, and processes to include authorizations and "need to know" factors for individual users using Attribute-Based Access Control (ABAC) (Priebe 2007) or Risk-Adaptive Access Control (RAdAC) (McGraw 2014) can be managed by the Enterprises by proving user identities, certificates, passwords, user roles, and user locations.

Unlike other blockchain approaches, Philos™ is optimized to maximize five categories of trust, automate that process, and use dashboards to help monitor and control trust enterprise levels. The foundational cybersecurity goals of CIA are automatically monitored and maximized through the five categories: Suppression Trust, Validation Trust, Reliability Trust, Refutation Trust, and Deprivation Trust, as defined by Gorog and Boult in 2019 (Gorog and Boult 2019). The uniformity of trust methods across the enterprises should help enforce the uniformity of ZT applications throughout and provide an underlying support structure to support effective security. In addition, Philos™ also has a unique combination of designed-in capabilities: indefinite scalability, neutral governance, auditability, no mining or wasted energy, end-of-life planning, minimal transaction size, offline storage, individual data sales, and rapid parallel transactions.

MBSE and MBSAP are designed to deal with complex systems and can use ASE as an effective approach to dealing with dynamic changes. According to Borky and Herber (Borky and Herber 2020), "An MBSE foundation is indispensable if ASE is to be consistent, measurable, high quality, and efficient." To deal with methods to control Negative Emergence Complexity, measurability is particularly important to be able to monitor whether controls are effective. MBSE and MBSAP are therefore not only highly supportive and synergistic with architectures based on the ESP™, secure cryptographic key provisioning, and Philos™ products but appear to be the best way to determine when the total benefits of these three applications are greater than any of them working individually.

5.4 ECOSYSTEM SOLUTIONS

The proposed solutions to reduce behaviors of negative emergent complexity from malicious cybersecurity threats through embedded cybersecurity design, effective uniformity of defensive methods using BlockFrame®'s novel ESP™, secure cryptographic key provisioning, and the Philos™ blockchain Marketplace. These three products, along with effective systems engineering, will improve control structures and support DiD and ZTA. The novel BlockFrame® approach to systems engineering-based cybersecurity architectures is leading the marketplace with the first-of-a-kind third-generation Philos™ blockchain/distributed ledger concept for advanced logistics using secure cryptograph key provisioning for the IOT. The Philos™ blockchain is being designed using BlockFrame®-created inventions for the unique security management capabilities shown in Table 5.1 (Gorog and Lambert 2021). Many of the capabilities shown in Table 5.1 are also necessary to realize the system applications necessary to provide many cybersecurity services, IoT security, and network security capabilities discussed in this chapter. The fact that all three products are designed with indefinite global scalability in mind also makes them uniquely expandable. Risk identification, risk analysis, and risk reduction are integral to the systems engineering goals of measuring and reducing the financial costs of losses to cyberattacks. In other words, they will help produce an ecosystem-wide environment of positive emergence to control the recognized negative emergence of current cybersecurity threats. Positive emergence can be expected through the uniformity of actions and products designed to improve distributed system behaviors in the areas of balancing privacy and visibility, accountability, secure logistics, identification, traceability, and trust.

These ecosystem-wide technologies support several positive emergent behaviors and use cases, including the capabilities to protect sensitive government records, manage supply chains, and reduce the risks of many cybersecurity threats by using MFI to improve visibility, and to identify and prove trusted participants, components, and software. The solutions support the capability to implement practical applications of MBSE and MBSAP to complex cybersecurity systems and many use cases. Robust systems engineering will identify, replicate, and reuse successful design patterns and reference architectures to help quickly proliferate designs to broader sets of applications. BlockFrame, Inc., has unique subject matter expertise in these areas to provide consulting services and support to companies and government entities through systems engineering and risk management for their individual use cases and requirements.

One of many practical use cases to describe the application of these BlockFrame® technologies is to secure and manage energy grids for robust cybersecurity and management of transactive energy (Gorog and Lambert 2021). It is well known that modern energy grids are highly complex and high-risk targets for cyberattacks and ransom attacks of national concern. As observed in the vulnerability of the Colonial Pipeline ("Colonial Pipeline Cyber Incident" n.d.), executive orders from both President Trump (Donald Trump 2020) and President Biden ("Executive Order on Improving the Nation's Cybersecurity" 2021), joint reports by the FBI and the National Security Agency (FBI/NSA 2018), and many other examples, United States critical energy infrastructure is known to be under threat of foreign attack, and in many cases has already been penetrated. In 2021, the US Government more specifically identified cybersecurity shortfalls due to the lack of attention to Zero Trust capabilities ("Federal Agencies Face New Zero-Trust Cybersecurity Requirements

TABLE 5.1
Third-Generation Distributed Ledger

Generations of Blockchain and Capabilities	First Generation			Second Generation			Third Generation
	Bitcoin	Crypto Currency	Stable Coins	Ethereum	Hyper-Ledger	Stellar	Philos
No third party needed for transaction	×	×	×	×	×	×	×
Content publicly verifiable	×	×	×	×	×	×	×
Volatile speculation	×	×		×			
Publicly accessible	×	×	×	×			×
Cost effective			×			×	×
Operation of smart contracts	Smart operations and application			×			×
Control and vetting participants					×	×	×
Minimize data usage					×	×	×
Expandable to large scale					×	×	×
Separation of multiple industries	Governance cross-industry support and organizational management controls						×
Tracing trust of individuals							×
Separation of private data							×
Control of multiple jurisdictions							×
Licensing of transaction contracts							×
Economic indicator controls							×
Indefinite modular scalability							×
Zero-trust ledgers							×
No wasted energy consensus							×

The ×'s show the capabilities of each generation of blockchain

| CSO Online" n.d.) and planned to start a new office to advance the use of Zero Trust concepts ("Pentagon to Launch Zero Trust Cyber Office in December" n.d.). Even without complex analysis or modeling, these events represent strongly negative emergence from cybersecurity threats, and government reactions represent actions they hope will provide some degree of positive emergence to mitigate those threats. Unfortunately, the rapidity of penetration and destruction from cyberattacks does not yet inspire confidence that defenses are effective now or will be in the foreseeable future.

Technically, modern energy grids have highly complex routing and financial considerations with associated microgrids, net-metering requirements for alternative energy sharing with individual subscribers, requirements to maintain adequate baseloads during period of low availability and rapid switching of alternative energy sources, and political requirements for non-fossil fuel, non-nuclear, and non-hydro portfolios. These complex requirements are unprecedented with previous generations of power systems and suggest that unprecedented levels of smart digital support, cyber defense, and supervisory control and data acquisition (SCADA) capabilities will be required. All of these complex factors combine to create an unprecedented potential for negative emergence across the entire energy industry (Donald Trump 2020).

BlockFrame®'s solutions for transactive energy use cases are representative of capabilities that can be applied to many complex systems with high levels of negative emergence. BlockFrame® ESP™ devices can be used in an unlimited number of nodes and devices to collect system data and, in conjunction with blockchain management, immutably and non-reputably record and manage that data by producing transaction contracts using non-fungible tokens (NFTs) over a wide variety of human, hardware, or software actors and events. ESP™ devices can also be used to establish zero-trust rules and securely deliver and provision software and software updates at each node of interest in complex energy grids of producer, distribution, or consumer delivery points. Figures 5.2–5.5 show a decision path for a notional energy company with distributed energy resources (DERs) to deal with transactive energy decisions in a hostile cybersecurity environment. We assume that Third-Generation BlockFrame® technologies and procedures are available to address qualitative management issues and needed capabilities:

Capability 1: Provision robust cybersecurity for DERs using NFTs
Capability 2: Collect and manage immutable forensic proof of all selected energy transaction data from DERs.
Capability 3: Scale practical blockchain management for an entire energy grid.

Each of these capabilities are currently needed to manage future transactive energy resources but are currently not available. They also address current issues that represent negative emergence. Creating these capabilities to mitigate current problems can also represent a process of positive emergence.

To address Capability 1, BlockFrame proposes using NFT technologies to provide robust cybersecurity capabilities to secure DERs. Figure 5.2 (Gorog and Lambert 2021)

Blockchain for Distributed Energy Resources Security

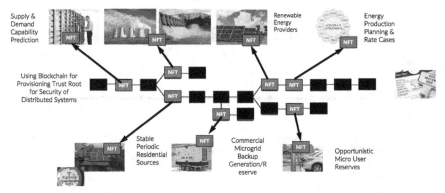

FIGURE 5.2   Using NFTs to provision security.

shows that the same NFT transaction contract tokenization can be used throughout energy enterprises, thereby reducing the risks of negative emergence. This NFT process through BlockFrame® ESP™ devices is approximately analogous to the definitions and functional behaviors of Emergent Behavior Observers (EBOs), as described in (Mittal and Rainey 2015) and (Zeigler 2019), with the addition of other security, encryption, provisioning, and recovery functions that allow behaviors that might be better described as Emergent Behavior Controllers (EBCs).

The challenge of proving valid energy flow, and therefore valid energy transactions, is shown in Figure 5.3 (Gorog and Lambert 2021). Both security and management can be extended to provide provable, reliable, and trusted management of financial data when delivered from various nodes of DERs, including energy exchanges, energy credits, and interactive consumer participation. This would typically involve cyber-physical ESP™ devices to establish hardened zero-trust processes, and then establish highly encrypted asynchronous communications through secure cloud services among all devices and their networks. Collected data would then be managed by the secure blockchain system, so all energy flow data and their associated costs would be indelibly recorded on the blockchain for accurate proof of valid transactions. Transaction contracts would also be non-reputable as their processes would record all necessary data to prove their validity.

High levels of security and provable data for customer billing are essential for the sustainability of viable energy systems. However, highly distributed systems must also be scalable to the appropriate size, locations, and the volume of data collected by modern transactional energy grids. Previous tests and studies of the scalability of blockchains (Gorog and Boult 2019) show that unless the blockchain is capable of storing massive volumes of transaction data, it is impractical to store those data on distributed servers, and still maintain the performance requirements of the blockchain itself. Those tests and operational requirements led to a bottom-up development of the Philos™ blockchain, which is based on low-cost, energy-efficient, and indefinitely scalable management of blockchain operations. The third-generation blockchain capabilities shown in Figure 5.4 (Gorog and Lambert 2021) demonstrate

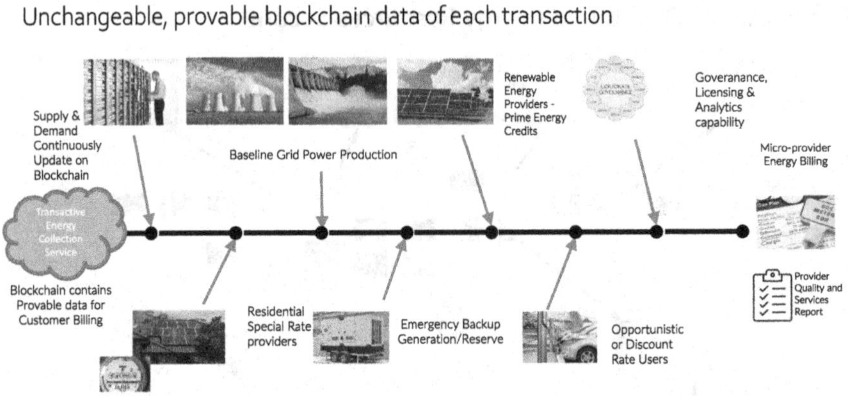

Unchangeable, provable blockchain data of each transaction

FIGURE 5.3 Collecting immutable and non-reputable data for customer billing.

Generations of BlockChain and Capabilities	First Generation			Second Generation			Third Generation
	Bitcoin	Crypto-currency	Stable Coins	Ethereum	Hyper-ledger	Stellar	Philos
No 3rd Party needed for Transaction	x	x	x	x	x	x	x
Content Publicly Verifiable	x	x	x	x	x	x	x
Volatile Speculation	x	x		x			
Publicly Accessible	x	x	x	x			x
Cost Effective			x			x	x
Operation of Smart Contracts				x	x	x	x
Control and Vetting Participants					x	x	x
Minimize Data Usage						x	x
Expandable To Large Scale					x	x	x
Separation of Multiple Industries							x
Tracing Trust of Individuals							x
Separation of Private Data							x
Control of Multiple Jurisdictions							x
Licensing of Transaction Contracts							x
Economic Indicator Controls							x
Indefinite Modular Scalability							x
Zero-Trust Ledgers							x
No Wasted Energy Consensus							x

FIGURE 5.4 Applying indefinitely scalable third-generation distributed ledgers to manage DRE.

BlockFrame®'s approach to solving those scalability problems, which are necessary for practical automated management of secure and effective financial management of complex distributed power grids.

Once established, networks using BlockFrame® technologies provide new management capabilities for energy providers, as represented in Figure 5.5 (Gorog and Lambert 2021). As energy producers must plan for co-mingling every "color" of power—and the corresponding "color" of its monetary value—from traditional and alternate forms of energy, from consumer-generated suppliers, and from other producers—they must be able to prove those details to government regulators, holders of energy credits, and consumers. With cryptographically secure collection devices,

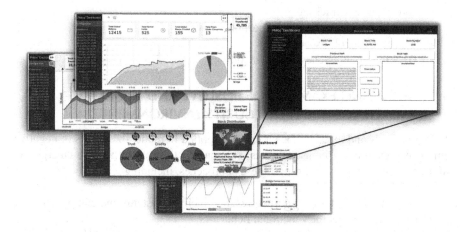

FIGURE 5.5   Applying indefinitely scalable third-generation distributed ledgers to manage DRE.

networks, and blockchains used for management and storage, these problems can be solved even in highly complex systems.

Because the Philos™ blockchain does not require expensive "data mining," it maintains very low energy-use and very low costs to operate. The combination of inexpensive NFT-based transactions, blockchain-based security management, and the ability to scale to widely distributed energy grids can streamline automated management, provide essential services at low costs, and save money for energy suppliers. Some of the resulting benefits of these new approaches are:

- Apply blockchain and NFT technology for IoT security
- Carbon-neutral blockchain without wasted energy consumption
- Security and privacy are integrated industry wide
- NFT technology is used for individual monetization of every transaction
- Blockchain saves money for utilities and consumers
- Immutable and non-reputable transaction contracts are used for payment records
- Pre-purchase consumption as a service capability
- Automated billing using utility tokens
- NFTs to track types and categories of energy use credits
- Peer-to-peer trading among co-ops
- Track rate changes and DER energy transaction history
- Use ESP™ proof-of-origin capabilities to verify communications from DERs
- Create geographical maps of grid events and DER origins
- Create a view to see DER transactions vs. energy rate changes
- Quality and effectiveness reports, and customer trends and patterns
- Reports and accurate billing for consumers
- Trend analyses
- Accurate rate cases

- Forensic usage tracking
- Accurately track and manage carbon credits
- Easy tracking of reserve access
- Tracking of emergency power shedding

5.5 MODELING EXAMPLES OF POSITIVE AND NEGATIVE EMERGENCE IN CYBERSECURITY ENVIRONMENTS

MBSE analysis and tools can be effectively applied to model the effects of negative emergence from cybersecurity threats and controls to produce positive emergence. Many benefits are derived through these approaches, including:

- Visualizations of SoS architectures, and their subsystems and components
- Definition of specific functional behaviors
- Verification of behaviors with engineers, programmers, and SMEs
- Validation of functional processes by designers, stakeholders, investors, and SMEs
- Replications of reference architectures for reuse and more rapid development
- Discovery of system flaws and vulnerabilities
- Identification of points and processes of risk, calculation of risks, and risk mitigation

For this chapter, the United States Naval Postgraduate School (NPS) Monterey Phoenix (MP) tool MP-Gryphon (Auguston 2009a b) is used to demonstrate examples of negative and positive emergent behaviors in a selected cybersecurity framework. The use of MP can also be compared to other approaches in this book. Several assumptions will be made for the purposes of the scenarios (Schema) being modeled:

- Monterey Phoenix is designed and used as a minimalist model that is not intended to model events or behaviors at refined levels of detail, such as all engineering processes within system of systems, but is more useful to display an exhaustive number of high-level behavioral outcomes for examination by analysts, engineers, and stakeholders to qualitatively verify and validate those behaviors.
- While some "simulation" is possible, a relatively small number of iterations (or "Scope") are usually used to allow visualization for the user, which allows users to make relatively rapid changes to the modeled behaviors while still providing a scope-complete set of behaviors. This MP design and operation are based on the Small Scope Hypothesis (Jackson 2012), which assumes that "most flaws in models can be demonstrated on small counter-examples" (Giammarco and Auguston 2019). The relative simplicity of this approach is intended to discover "unknown unknowns" through rapid visualizations that can be communicated with verification and validation groups to discover design flaws or opportunities for process improvement.
- MP can be used as a commonly accepted approach for basic emergence research. The software is jointly maintained by NPS and the National Security

Agency (NSA) and is gaining exposure and acceptance within Department of Defense modeling communities. While it is not intended to provide a comprehensive MBSE model, it is a useful behavioral analysis tool that if used consistently will provide ontologies and processes that can be transferrable and understandable within diverse academic and topical settings.

- Emergence theory is inherently about chaotic environments where there may be little information about actual causes and effects. As such, providing high levels of resources in attempts to model high levels of detail can be both futile and expensive because the subject is known to be chaotic outside the realm of methods to control it, or even to understand it in detail. Because of the relatively simplified Monterey Phoenix focus on logical events and behaviors, it is more of a "what if" approach to validate functions at a relatively high level to allow analysts and stakeholders to visualize high-level trends and effects, and to examine and "think through" alternatives fairly rapidly. Typically, MP will be used to examine trees of logical alternatives, but once defined, the model will automatically and exhaustively (up to the Scope set by the user) produce event trees (traces) that can be examined for unexpected outcomes. In short, the MP approach to examine emergence is to logically discover and help SMEs think through disconnects that might cause unexpected or strong emergence, and attempt to mitigate negative emergence.
- This chapter will assume that a very simplified representation of a generic blockchain process can be used as a surrogate for more advanced systems that might be vulnerable to cyberattacks. This might be only one of many such analyses for complex systems like transactive energy enterprises. For analysis purposes, we will examine hypothetical situations with high risk and low levels of cyber security to help a method to identify subjective levels of negative and positive emergence and risk. This approach does not attempt to replicate the high levels of capability in advanced blockchains like Philos™ but is purposely focused to examine cyber risks as negative emergence and the need for advanced cyber protections (as positive emergence controls). More detailed models of actual blockchains are not included for several reasons, including possible exposure of proprietary processes, unintended exposure of possible security flaws, or possible liabilities of unfair comparisons with actual products.
- Monterey Phoenix has some capabilities for data generation and analysis, and those are in the process of being expanded. For this chapter, MP will not be used for parametric sensitivity analysis. Selected MP behavioral diagrams (traces) will be examined, but hypothetical compound probability analysis and parametric sensitivities will be calculated and displayed graphically using Microsoft Excel for convenience.

Monterey Phoenix v4 Gryphon GUI v1.0.0 (MP-Gryphon) was used to produce behavioral diagrams (traces) to model a very generic blockchain process to demonstrate normal blockchain operations when effective cybersecurity is assumed to be in place, and then to exhaustively demonstrate events where attacks might occur if defenses fail. At Scope = 1, one trace was generated, representing one pass through

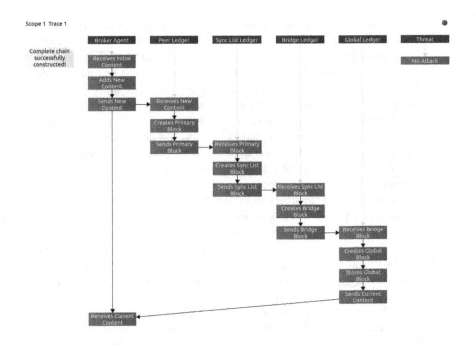

FIGURE 5.6   Successful build of blockchain on the first cycle with no successful cyberattacks.

the blockchain build cycle without successful threats. That example process is shown in Figure 5.6. In simple terms, the process starts by collecting data that users want to add to and store on the blockchain. Cooperative methods merge data from each user and combine those data with data from many other users, sometimes with high levels of separation of distance, function, management, jurisdiction, and security. At the end of the process, all the data from the many users will merge (through consensus decisions by the users) into a new block which is added to all the previous historical blocks in the blockchain.

At Scope = 2, Monterey Phoenix automatically generated 16 traces, representing two blockchain build cycles. Of the 16 traces, 2 represented that the blockchain was successfully constructed, and 14 represented successful cyberattacks if defenses failed at particular events. Figure 5.7 represents two passes through the blockchain build cycle without successful threats, while Figure 5.8 represents a failure of cyber defenses while a Sync List Ledger is attempting to build a new Sync List Block during the second cycle of a blockchain build process.

At Scope = 3, 44 traces were generated, representing three blockchain build cycles. Of the 44 traces, 3 represented the blockchain as being successfully constructed, and 41 represented successful cyberattacks if defenses failed at particular events. Figure 5.9 represents three passes through the blockchain build cycle without successful threats, while Figure 5.10 represents a failure of cyber defenses while a Broker Agent is attempting to Add Content to New Block.

In all, Monterey Phoenix may generate hundreds or thousands of event traces representing different behaviors, depending on the logical variations of interest to

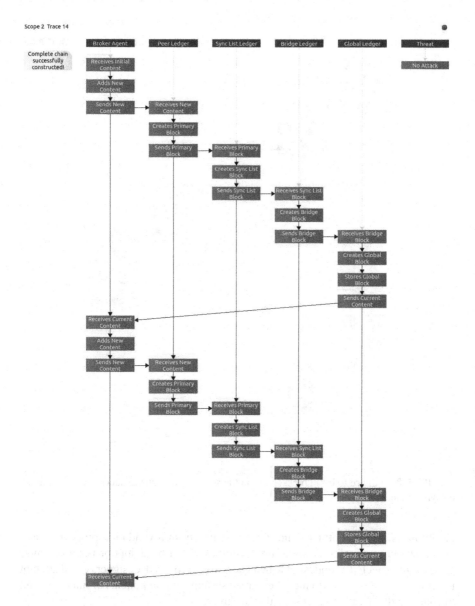

FIGURE 5.7 Successful build of blockchain through the second cycle with no successful cyberattacks.

analysts, but the concept of the Small Scope Hypothesis suggests that examination of that many alternatives is probably past a point of diminishing returns and could be a waste of time for knowledgeable analysts to discover whether or not significant negative or positive emergence is observable. When trying to discover and mitigate negative emergence in complex chaotic systems, this model-assisted analytic process could be iterative to discover and correct a variety of factors like unexpected

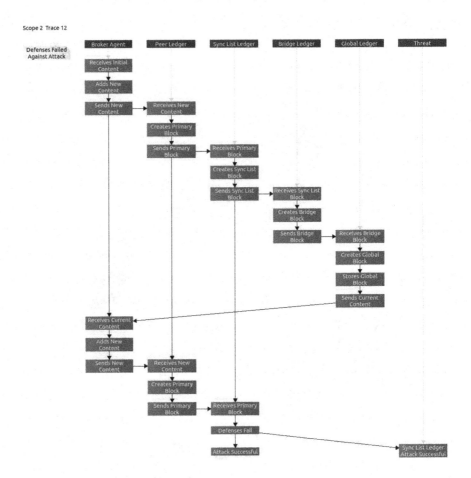

FIGURE 5.8   Failure to build blockchain on the second cycle due to successful cyberattack on sync list ledger.

design flaws, external attacks, unexpected results from a validation process, or any "unknown unknowns" that might produce unexpected behaviors or negative emergence. If unexpected or undesirable behaviors are discovered, designers could attempt to mitigate them or move to model other subsystems or components of interest—and to continue the iterative process as they desire.

Just such an example occurred when an early version of this MP model of this Schema showed a Broker Agent being attacked during its construction of the initial block—shown in Figure 5.11. Instead of halting the construction, the logical model of the "expected" behavior sent corrupted data on to the next step in the blockchain process. While this was discovered to be due to a logical error in the initial modeling process of the sequence of behaviors, it could represent the many kinds of errors that sometimes do occur in real-life design due to unintentional programming errors, inexperienced programmers writing poor quality "spaghetti code," misunderstandings between development teams, the acceptance of unvetted off-the-shelf code, or

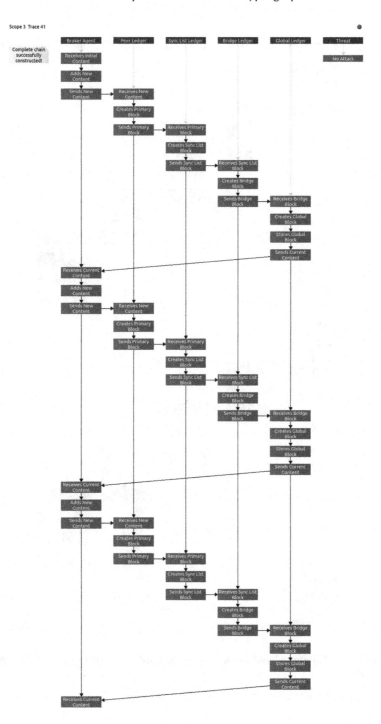

FIGURE 5.9   Successful build of blockchain through the third cycle with no successful cyberattacks.

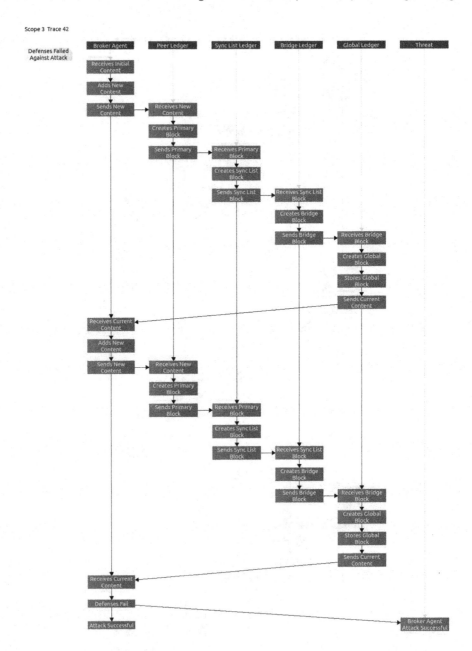

FIGURE 5.10 Failure to build blockchain on the third cycle due to successful cyberattack on broker agent.

FIGURE 5.11   Successful cyberattack on broker agent where MP diagram discovered emergent behavior in system design.

errors in development/security/operational (DEVSECOPS) design itself. Paraphrasing Murphy's Law, "If it could happen, it will happen!" In this case, the impact of that little error could have been the disastrous compromise of the core function of any blockchain, the perfect proof of mathematical security. Verification of behaviors like this is exactly how Monterey Phoenix itself identified unexpected emergence while writing this chapter, exposed that behavior, and allowed it to be corrected.

Many other processes could be modeled for the same purpose to identify emergence, but because of the multilayer architecture of blockchains, this Schema was selected to show how a system of systems will have various layers of attack surfaces that are vulnerable over time. Successful mitigation of risk must consider an adequate sampling or potential risk events over time and introduce effective procedures that will lower risks to a manageable level. When cyber security risks (as forms of negative emergence) are reduced to the point they are understandable and controllable, emergence could be considered relatively more positive. The scenario of an attack on a Broker Agent at Figures 5.10 or 5.11 may be of particular interest to analysts as a likely point of attack because it is might be the most exposed to external users through application programming interfaces (APIs) and involves more human factors that are open to compromise or error. Although other points of attack are possible, they are probably less likely because they are not as exposed and are totally automatic machine-to-machine interfaces, and the data will be encrypted throughout the block-building process if a particular blockchain (like Philos™) is designed to do that. Most blockchains are not designed to encrypt data in transit but may leave source data exposed throughout their life cycle. As shown in Table 5.1, many of those shortfalls are recognized and mitigated through the design of the Philos™ Blockchain. In the Monterey Phoenix example and Schema and scenarios shown here, we assume that all Philos™ capabilities could be included in the model and that "near perfect security" could be attainable. In this simple example, if analysts are satisfied that the threats shown in Figures 5.10 and 5.11 are in fact more interesting than other traces, they have the option of studying them in more depth, expanding the simulation beyond two or three iterations to many iterations, testing new behaviors like new cybersecurity capabilities, or conducting a variety of other logical tests.

5.6   QUANTIFYING AND PRIORITIZING CYBERSECURITY RESPONSES

Once a particular attack point is identified and its characteristics and environment are at least partially known, analysts may use simple methods to quantify and compare the relative effects of interest. In the example in Figure 5.10, if a relatively high potential point of vulnerability is discovered, analysts might want to measure the potential risk of that vulnerability. In a highly chaotic environment, attempting to simulate complex cyberattacks in detail might not be worth the effort if simple calculations are able to answer the question of whether defenses are robust enough to contain or mitigate the complexity and costs attackers might incur, or whether users think that their defenses are "robust enough" to mitigate their risk to a manageable level. Referring back to Figure 5.1, that might be described as moving from out-of-control "strong" or "spooky" negative emergence to a point where emergence complexity can be controlled within the "knowledge boundary" where known methods of mitigation can be used.

If quantitative probabilities of attack success can be estimated, based on engineering, historical, or even anecdotal examples, the magnitude of a probability of survival of the target of the attack could be calculated using the formula:

$$P_{\text{survival}} = \left(1 - P_{\text{kill}}\right)^N \qquad (5.1)$$

where the survival of any system or component is simply $1 - P_{\text{kill}}$ for each attack. As a simple example, if the probability of "kill" from a single attack attempt is 0.05, the probability of "survival" from that single attack is $(1 - 0.05)$ or 95%. However, if multiple attacks were attempted at the same probability of kill, the probability of the target surviving may rapidly diminish. At $n = 10$ attacks, or $n = 100$ attacks, the probability of the target surviving falls to about 0.60 and 0.006, respectively.

For a cybersecurity example, a "kill" could be from many mechanisms, including denial of service, physical destruction, theft, or a successful ransom of data. Some "kills" might also be undetectable to users, which is the basis of many reports of cyberattacks that may conclude that the target was under attack for 9 months and never knew it. Whatever the mechanism, SMEs might have to make some "well-informed guesses" as to the potential relative risk of attacks and quantify the relative vulnerabilities of system components over time. In an environment where cyberattackers have diverse tools and methods to attack over time, they also force defenders into requirements to maintain almost perfect defenses in order to sustain themselves over a long period of time. However, if defenders can identify their most likely relative vulnerable points of procedures, they should also be able to make their defenses more robust over time. For example, if a defender can apply techniques to lower the attacker's single attack probability of kill to 0.00001, even after 100 attacks, the probability that the target might survive is 0.99900.

5.7 APPLYING MONTEREY PHOENIX TO QUALITATIVE CYBERSECURITY LOGIC

Although attaining a perfect defense is theoretically impossible, a robust unified zero trust defense should strive to attain a low enough risk level over time as a quality requirement for all system components, which could deter attackers from expending their time and cost to attack in the first place. The most important consideration is that risk might not have to be reduced to zero if defenses are adequate to make attacks too expensive or too vulnerable to counterattacks. Admittedly, those decision factors are somewhat subjective and qualitative, but they might be sufficient with decision-support models like Monterey Phoenix to help defenders examine how much effort is enough to defeat the expenditures and threats made by their attackers. The use of cryptographic provisioning devices like the ESP™ and blockchain systems like Philos™ that are designed to monitor and improve the quality of trust factors used by individual users can regulate that quality by peer pressure from other users through incentives for quality compliance over time. In combination with modeling behaviors within system architectures, Monterey Phoenix could also be used to model behaviors of defenders to compare cybersecurity threats and defensive strategies over time. Without robust defenses, victims of cyberattacks could experience the effects of attacks over extended periods of time, might not know they are being attacked, and suffer effects that include data breaches, loss of customer data, penetration into core company functions, disruption or permanent destruction of data or operations, monetary losses, or ransom attacks. Using models generated from Monterey Phoenix to examine and think through those problems should help identify and control many destructive and urgent problems first, which might not be practical even with larger and more expensive approaches.

5.8 SUMMARY

This chapter addresses the concepts of negative and positive emergence with practical examples of digital operations within hostile cybersecurity environments.

Cybersecurity has been shown to be highly chaotic and represents cases of negative emergence that impede normal and expected operations beyond the point of subject matter experts' abilities to apply known controls. BlockFrame, Inc., has invented and designed technologies and procedures that are specifically intended to help control the negative emergence of malicious cyber threats and to add controls that can result in positive emergence. These processes are first explained theoretically and then demonstrated in one practical example of the requirements of a notional energy company. The usefulness of MBSE is demonstrated by examples using the Naval Postgraduate School Monterey Phoenix tool to show normal operations of generic blockchain design behaviors and then compares several possible points of attack by generic cyberattack behaviors. Examples of relatively simple scenarios are compared using the Small Scope Hypothesis to demonstrate how behavioral modeling can be useful to support decisions to think through needed cyber defenses and capabilities, as well as assisting in model verification and validation and the discovery of unexpected emergent behaviors.

REFERENCES

Auguston, Mikhail. 2009a. "Software Architecture Built from Behavior Models." *ACM SIGSOFT Software Engineering Notes* 34 (5): 1–15.

Auguston, Mikhail. 2009b. "Monterey Phoenix, or How to Make Software Architecture Executable." Presented at the OOPSLA '09/Onward Conference, OOPSLA Companion, pp. 1031–1038, October.

Borky, John M., and Thomas H. Bradley. 2019. *Effective Model-Based Systems Engineering.* 1st ed. Cham: Springer.

Borky, John M., and Daniel R. Herber. 2020. *Engr 667: Advanced Model-Based Systems Engineering; Session 13: Cybersecurity of System Architecture (Lecture Slides).* Fort Collins, CO: Colorado State University.

"Colonial Pipeline Cyber Incident." n.d. Energy.Gov. Accessed November 27, 2021. https://www.energy.gov/ceser/colonial-pipeline-cyber-incident.

Commission on Enhancing National Cybersecurity. 2016. "Report on Security and Growing the Digital Economy."

"Cyberspace Solarium Commission." n.d. United States of America. https://www.solarium.gov/.

"Domain Name System." n.d. Wikipedia. Accessed May 18, 2020. https://en.wikipedia.org/wiki/Domain_Name_System.

"Executive Order on Improving the Nation's Cybersecurity." 2021. The White House. May 12, 2021. https://www.whitehouse.gov/briefing-room/presidential-actions/2021/05/12/executive-order-on-improving-the-nations-cybersecurity/.

FBI/NSA. 2018. "Alert (TA18–074A) Russian Government Cyber Activity Targeting Energy and Other Critical Infrastructure Sectors." https://www.us-cert.gov/ncas/alerts/TA18-074A.

"Federal Agencies Face New Zero-Trust Cybersecurity Requirements | CSO Online." n.d. Accessed November 27, 2021. https://www-csoonline-com.cdn.ampproject.org/c/s/www.csoonline.com/article/3632878/federal-agencies-face-new-zero-trust-cybersecurity-requirements.amp.html.

Giammarco, Kristin, and Mikhail Auguston. 2019. "Monterey Phoenix Tutorial." May.

Gorog, Christopher. 2019. "BlockChain Cybersecurity Logistics Organization for Globally Scalable Security Management." Unpublished Draft.

Gorog, Christopher, and Terrance E. Boult. 2019. "Solving Global Cybersecurity Problems by Connecting Trust Using Blockchain." University of Colorado -- Colorado Springs.

Gorog, Christopher, and Kent Lambert. 2020. "National Security Products, Cryptology, Trust, and Blockchain Frameworks to Secure Supply Chains and Operations."

Gorog, Christopher, and Kent Lambert. 2021. "BlockFrame, Inc., Technology, Blockchain for Transactive Energy Solutions." July.

Jackson, Daniel. 2012. *Software Abstractions: Logic, Language, and Analysis*. Cambridge, MA: MIT Press.

Lambert, Kent D., Angela Williams, Joann Ginal, and Bob Rankin. 2018. "Colorado Senate Bill SB-086, 'Cyber Coding Cryptology for State Records.'" Colorado General Assembly. http://leg.colorado.gov/bill-search?search_api_views_fulltext=sb18-086.

McGraw, Robert. 2014. "Risk Adaptable Access Control (RAdAC)." https://csrc.nist.gov/csrc/media/events/privilege-management-workshop/documents/radac-paper0001.pdf.

Mittal, Saurabh, and Larry Rainey. 2015. "Harnessing Emergence: The Control and Design of Emergent Behavior in System of Systems Engineering." In *Proceedings of the Conference on Summer Computer Simulation*. Chicago, IL: Society of Computer Simulation International.

National Cyber Security Centre. 2019. "The Cyber Threat to UK Business 2017–2018 Report." United Kingdom. https://enterprise.verizon.com/resources/reports/2019-data-breach-investigations-report.pdf.

Office of Personnel Management Security Resource Center. n.d. "Cybersecurity Incidents." Office of Personnel Management. https://www.opm.gov/cybersecurity/cybersecurity-incidents/.

Paloalto Networks. n.d. "What Is Zero Trust?" Accessed May 18, 2020. https://www.paloaltonetworks.com/cyberpedia/what-is-a-zero-trust-architecture.

"Patents Filed by BlockFrame, Inc. U.S. Patent Application No. 17/488,529; No. 17/488,562; No. 17/488,589; No. 17/488,655; Patent Cooperative Treaty (PCT) Application No. PCT/US21/52839; No. PCT/US21/52854; No. PCT/US21/52865; No. PCT/US21/52872." 2021. United States Patent and Trademark Office.

"Pentagon to Launch Zero Trust Cyber Office in December." n.d. Accessed November 27, 2021. https://www.securitymagazine.com/articles/96504-pentagon-to-launch-zero-trust-cyber-office-in-december.

Pratt, Mary. 2018. "What Is Zero Trust? A Model for More Effective Security." CSO United States. https://www.csoonline.com/article/3247848/what-is-zero-trust-a-model-for-more-effective-security.html.

Priebe, Torsten, Wolfgang Dobmeier, Christian Schläger, and Nora Kamprath. 2007. "Supporting Attribute-Based Access Control in Authorization and Authentication Infrastructures with Ontologies." *Journal of Software: JSW* 2 (1): 27–38.

Rainey, Larry B., and Andreas Tolk. 2015. *Modeling and Simulation Support for System of Systems Engineering Applications*. Hoboken, NJ: John Wiley and Sons.

Trump, Donald. 2017. "Presidential Executive Order on Strengthening the Cybersecurity of Federal Networks and Critical Infrastructure." The White House. https://www.whitehouse.gov/presidential-actions/presidential-executive-order-strengthening-cybersecurity-federal-networks-critical-infrastructure/.

Trump, Donald. 2020. "Executive Order on Securing the United States Bulk-Power System." White House. https://www.whitehouse.gov/presidential-actions/executive-order-securing-united-states-bulk-power-system/.

Utermohlen, Karl. 2019. "Facebook Data Breach 2019: 540 Million Users' Records Exposed." InvestorPlace. https://investorplace.com/2019/04/facebook-data-breach–2019/.

Verizon. 2019. "2019 Data Breach Investigations Report." https://enterprise.verizon.com/resources/reports/2019-data-breach-investigations-report.pdf.

Zeigler, Bernard P. 2019. "DEVS-Based Modeling and Simulation Framework for Emergence in Systems of Systems." In Larry Rainey (ed.), Engineering Emergence, A Modeling and Simulation Approach. Boca Raton: CRC Press.

6 The Nature of Emergence in Wargaming
Extracting Operational Insight from a Dynamic Game Field

William J. Lademan
Marine Corps Warfighting Laboratory

O. Thomas Holland
Georgia Tech Research Institutue

CONTENTS

> What good is experience if you do not reflect?
>
> *Frederick the Great*

6.1   INTRODUCTION

The universe operates in a manner that precludes absolute knowledge. However, local precisions exist and in sufficient amounts to permit confidence in action and successful outcomes. Thus, I can know enough to act; I can't know enough to be certain. Wargaming is a way of organizing experience and relating to reality and is a collaborative and creative method of conducting exploration, discovery, and synthesis. The

DOI: 10.1201/9781003160816-8

hallmark of wargaming is the incorporation of the human intellect (subjective, yes, but designing, penetrating, and embracing) into a methodology that produces intelligibility in the context of an emerging reality. By enabling and directing this human strength, wargaming thereby interprets a dynamic environment, bounds solution space, proposes relevant insight, and causes the emergence of governing factors. Wargaming accomplishes this as a system of systems which includes a scenario, design, methodology, adjudication, and assessment systems. All these complement, sustain, and enable the most dynamic and powerful system of all, the human intellect, in the search for discerning resolutions to problems contained in uncertain environments.

6.2   BACKGROUND

Hidden jeopardy resides in any attempt to exploit an advantage you don't possess.

In 1754, as a result of French incursions in the Ohio Valley, the British government determined to send an expeditionary force under General Edward Braddock to seize Fort Duquesne and resolve the territorial dispute. General Braddock's orders contained the following sentence:

> The very name of an expedition implies risk, hazard, precarious warfare, and a critical condition requiring quick resolves and rapid execution.

> *(Orders in Council, Expedition in Virginia 1974)*

This remark could reside comfortably in any Service's testimony before Congress.

The resulting campaign was a model of 18th-century planning and execution and a precedent for modern joint and combined operations. A task-organized force was assembled in and transported from Britain. The force was augmented with Provincials, Indian allies, and a Royal Navy attachment after arrival in Virginia. Supplies, wagons, animals, foodstuff, an artillery train, and ammunition were contracted for and assembled albeit with some difficulty. Intelligence was gathered; camp sites were selected along the route; a road was famously engineered and constructed through the wilderness; an order of march and operational schedule were formulated. Even the French were complementary when considering the level of organizational expertise displayed in assembling and operating the force. However, the advance ended in disaster a few miles short of the objective when, on 9 July 1756, the expeditionary force was attacked in a fluid double envelopment and destroyed as an organization. Why?

Among many reasons, one emerges as deciding the outcome. General Braddock envisioned the expedition as ending in a formal siege of a fortress occupied by the French and Indians and bound by the confluence of two rivers. He considered and planned for no other possibility. Yes, his force was organized to deflect probes and withstand raids. But, his only concession to other possibilities was his reliance upon linear tactics, a style of warfare incapable of delivering a favorable decision in the environment his force was to fight in. The expedition failed because General Braddock determined on a specific end, configured a force to obtain that end, and allowed for no other eventualities. What General Braddock could not conceive of was his enemy's first thought. He diligently established the conditions for a battle his enemy never intended to fight. His orders notwithstanding, thinking about expeditionary warfare proved harder than organizing for it.

Even with the modern complexities of multi-domain, information centric, time constrained, distributed, and hybrid warfare, the hard lesson imparted some 270 years ago at the Battle of the Monongahela is still relevant. Advanced weapons, information management, and computational power are some of the modern elements of warfare that must be combined and arranged to both set conditions and achieve effects in an expeditionary campaign. But, the most exquisite synthesis of all these elements, while necessary, is not sufficient to guarantee success. This is so precisely because as an operation is designed and put into motion, the enemy does the same. Contact between these opposing schemes generates unpredictability, uncertainty, and chaos. Thus, system capability, capacity, and performance are not sufficient indicators in the measurement of success. The key to the problem of establishing the conditions for operational success is preparation, and the key to preparation is anticipation. As General Braddock learned, the danger not considered is the most dangerous and precludes learning and agility.

The strength of wargaming is the ability to consider a real problem in an anticipatory manner. This allows for the evaluation of tasks, risk, and outcome in the context of a projected operational environment so as to support design, contingency, art, and the emergence of governing factors.

6.3 DEFINITION

A wargame represents a band of reality created by the assembly and organization of variables that establishes a game field, supports dynamic decisions and adjudications, and facilitates the emergence of insights and solution space. In a wargame, players extract and exploit information from the game field which contains all the elements which constitute the operational environment. This information then acts as a substrate which supports decision-making leading to coherent, relevant, and effective actions in the production of wargaming consequences and end states.

Thus, a wargame is an artificial construct that replicates conflict and permits the human intellect to consider the real problems of how to act successfully in a hostile environment.

The art of the wargame is the design of an "artificial construct" that supports the consideration of a "real problem." It is this connection between "artificial" and "real" that supports the representation in the game of the dynamic and variable interactions which drive the problem to be considered. The science of a wargame is the means used that "replicates conflict" and supports the consideration of "how to act successfully in a hostile environment." This entails the correct identification, assembly, and value of the selected variables which interact to generate conflict in a plausible operating environment. Finally, it is the "human intellect" which acts as a point of synthesis and fusion in the wargame dynamic. The intellect engages this artificial portrayal of reality in a fashion that permits the resolution of the chaos, non-linearity, complexity, and catastrophe attendant in wargames and can fashion appropriate action perhaps supported by a computational process. This action of the intellect upon the dynamic of the game is what causes emergent behavior. Figure 6.1 represents the components and relationships in the construct of a wargame.

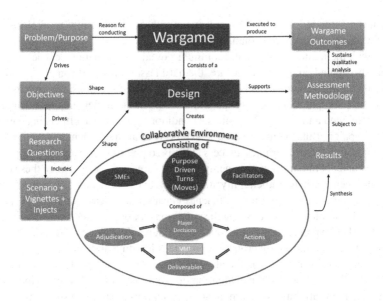

FIGURE 6.1 Wargame components and relationships.

The point of the above figure is not to detail and explain the specific steps, relationships, and components involved in wargame construction, execution, and assessment. Instead, it is to display the process by which data is converted into information,
that information is combined and arranged in the support of decisions, and those
decisions operate upon an environment to cause a transformed situation to emerge.
This emergence is caused by an action precipitated by a decision operating on the
variables contained in the game field. These variables interact to transform a situation
by causing one state of the variables to convert to a subsequent state. The emergent
state is informed by the resultant action and intent of the decision and the unintended
consequences which accompany it.

A wargame consists of the effort to translate a problem into gameable objectives,
supporting research questions, a design methodology that creates and supports a collaborative gaming environment aided by the selected use of methods, models, and
tools (MMT), and employing an assessment framework which delivers outcomes and
recommendations in a useful report format. The utility and power of this construct
lie in the integrated relationship among well-defined objectives, functional design,
guided execution, and emerging wargame outcomes.

Recent DoD-level interest has highlighted the importance and value of wargaming as a vital element in the comprehensive understanding of operational environments, force design, and operating concepts. There is both clarity and confusion in
this interest. Clarity in that it recognizes the problem of rapidly evolving adversaries
and gives momentum to an ability to anticipate and outpace these adversaries; confusion in that it has not distinguished between the natures of computational analysis
and wargaming. It has even been suggested that reinvigorating wargaming is merely
a matter of incorporating computational analytical techniques into wargame designs.

However, the effort to optimize this relationship entails more than the simple incorporation of MMT into wargame design. This incorporation is not new. During the period of the First Gulf War, wargames supported by models were conducted to successfully consider targeting, schemes of maneuver, and the matrix of strengths, weaknesses, threats, and opportunities. The struggle is to synchronize the process of sophisticated computational analytical methodologies with the action of the human intellect such that the potential of both are integrated and optimized by using the computational result as a substrate for human decision-making.

Computational analysis is based upon mathematical process; wargaming is based upon human judgment. Both are powerful and are compatible. But, they are not different expressions of the same thing. For its manipulation of data and its precision of result, computational analysis relies upon a methodology involving the quantification of variables and the specification of their interactions. In this analysis, computationally substantiated conclusions emerge from the connection of method to a specific problem. However, computational analysis is limited by the very tenants of its science to what is measurable. It cannot go beyond statements of trends and precision (accuracy is another matter) because it cannot substantiate what it cannot measure. Further, a particular resulting measurement does not necessarily imply a universal solution.

Wargaming rests upon what cannot be measured. This stands in contrast to but not in opposition to the computational analytical approach. A wargame does this by embracing, assembling, and organizing many variables without an attempt to assign values or calculate interactions. These variables, which reside in the situation, the individual, and emerge in the dynamic action of play, are impossible to measure in assembly or interaction. The action of the wargame generates interactions and relationships that could not have been anticipated and relies upon the emergence of results not subject to predetermination. All of this is synthesized and organized in the human imagination, and no science is capable of quantifying the path, dynamic, or chance that transforms this complexity into a comprehensible and coherent whole. And yet this is what both drives a game and defines its results. Thus, a wargame employs a coalescence of fluid variables and impossible to predict interactions to engineer an emergence of outcomes (think of a disordered pile of bricks, the variables and interactions, becoming a finished house, the outcomes (Ablowitz 1939)), which propose insights and direction. What then emerges are informed insights into specific problems guided by the following wargaming attributes:

a. **Purpose:** Accommodate the human intellect's agility and imagination in problem solving and innovation.
b. **Method:** Create a collaborative gaming environment that supports cycles of situational estimates, decisions, and adjudication.
c. **End State:** Propose the conceptual, technological, and operational basis for success in various operating environments.

Thus, wargames explore the interlocking coherence of the whole while computational analysis produces precision in isolation. The question is: How to associate the two to mutual benefit? The problem is one of relating processed facts and human imagination. The analyst and the wargame designer must combine the two realms without losing the essential strength of either in the midst of the constant dynamic and change in game play. The answer to this dilemma involves the recognition of the distinct natures of the two approaches and the effort to forge complementary methods. Wargaming permits judgment to be influenced in a dynamic context as evidence emerges as a precursor to decision. Analysis can aid this process by injecting points of precision into play which then merge with and act as an informing substrate for decision ... the universal requirement in any wargame. In other words, analytical methods can inform imagination with a precision designed to influence but not direct decision in game play.

The benefit of a wargame supported by analytical methods that provide points of departure and situational precision as the basis for decision is the production of informed and defensible insights that can shape and direct subsequent efforts in a concept or combat development sequence. There is no analytical methodology by which the outcomes of the inherently human activity of play can be transitioned into a rigid accuracy. But then war is an inherently human activity that only rarely adheres to the requirements of scientific law and rigidity in war rarely produces brilliance or success. The key to understanding the benefit of the incorporation of analytical methods into wargaming is that, while sharper and more focused insight can be expected as outcomes, one learns that knowledge does not have to be quantifiable in order to be defensible. This informed combination of the analytic science of the necessary with the wargaming art of the possible promises to provide a foundation for the objective substantiation and justification of what emerges from a wargame which can then inform and shape the resources and programs required for future military success. If the product of a wargame as a system of systems is emergence, then it becomes essential to understand what emergence is in the context of a wargame.

6.4   EMERGENCE

If Maier's definition of emergent behavior is accepted that,

> The system performs functions and carries out purposes that do not reside in any component system. These behaviors are emergent properties of the entire system of systems and cannot be localized to any component's system. The principal purposes of the system-of-systems are fulfilled by these behaviors.
>
> *(Maier 1998)*

then it can be extended by Ablowitz's thought as being representative of not simply an emergent realization of purpose through combined functions but also as a "new quality of existence that results from the structural relationship of its component parts" (Ablowitz 1939; Holland 2018). This extension is important because it allows an emergent property to transcend the limits of purpose and function and, given desired end states, actually transform an environment to advantage through the imperative of human concept and action.

Emergence is a sophisticated development of the idea that the whole is greater than the sum of its parts. Given what proceeds this chapter, there is no great need to review the theory of or consider the mathematical definitions relevant to emergent behavior in systems. Suffice it to say that among the factors which drive emergence—complexity, variety, constraint, and entropy—it is entropy that imparts the greatest effect in a wargame. If entropy is the measure of disorder in a system and if this disorder is expressed in the uncertainty that exists in the interpretation of a problem, in the representation of the continually adjudicated state of the game, and in the complexity of the outcome, then emergence in a wargame is the result of its structured mechanisms producing unpredictable outcomes. Here, we follow Pagels (Pagels 1989) in identifying the distinction in complexity between that which is structurally and that which is interactively complex ... structurally complex systems produce generally predictive behavior while interactively complex systems possess a lack of structure and a freedom of action which produce unpredictable behavior. Wargames are certainly interactively complex simply because their component systems, while designed to function predictably, are driven by the human intellect and have tendency to impose a richness and volatility in interaction that can produce an unanticipated and disproportional emergent behavior.

Wargaming is then a method which embraces the flux and turmoil of a reality devoid of absolute certainty in order to investigate the dynamic nature of a problem. Wargaming rests upon an intellectual and situational fluidity which cannot be quantified and embraces too many interacting variables to produce the precision expected in a computational result. Indeed, wargaming acts upon what cannot be measured, incorporates variables that cannot be quantified, generates interactions and relationships which could not have been anticipated, and so relies upon the emergence of results which could not have been predicted. The wargame as a system of systems operates, directs, and facilitates the process to form emergent outcomes but does not preordain the substance of the emergent outcome.

Thus, wargaming can excel in examining fluid and complex operational situations resistant to computation in which the opposing ideas and wills of the adversaries have as much to do with outcomes as the sum and array of the systems each side brings to the fight. Therefore, a wargame employs a coalescence of fluid variables, impossible to predict interactions, and dynamics not subject to law in order to engineer an emergence of outcomes which propose insights and direction on the premise of human assessment. Within this construct, emergent behavior occurs as a result of a participant's interaction with the game field (Figure 6.2), which is a construct containing the wargame's pertinent variables and information.

A game field is the wargame designer's construct containing all the relevant variables and factors that pertain to a game's particular purposes and objectives. As a field, the components are interrelated and change can produce highly non-linear effects in the arrangement, emphasis, and value of those components. It is the combination of computational analysis supplied by models and the cognitive analysis of the participants applied to the game field (this process represented in the figure by the variable X being transformed into a function of X by a computational or cognitive analytical process) which produces the results necessary to act as a substrate (variables contained in the bracket) for players to deliberate and arrive at a decision which allows the players to successfully act in a changing and hostile environment. This ability to modify, shape, and exploit an operational environment is what constitutes the basis for emergence in

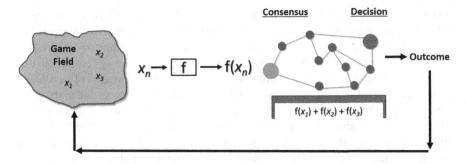

FIGURE 6.2 Game field supporting wargame decision process.

a wargame. The manifestation of this emergence is the ability to identify the governing factors, which represent the forces that drive functions and trends in an operating environment, and thus the identification of what is the key to further a successful engagement. Thus, emergent behavior is inseparable from and inherent in the play of a wargame. The knowledge of the existence of emergent behavior is not in question; its quality and applicability are. Therefore, the purpose of wargame art, science, and best practice is the creation of a design and an evaluation framework that facilitates and optimizes the production of inevitable emergent behavior and incorporates the methodology for evaluating that emergent behavior. Equation (6.1) is a representation of this complex relationship.

$$\varepsilon = S_N * E_N \qquad (6.1)$$

The equation relates emergent behavior (ε) to the final state of a wargame (S_N) acted upon by an evaluation process (E_N). While intermediate emergences are continual in the execution of a wargame, a cumulative emergence is produced in the identification of the final state of the wargame and the application of an assessment framework to that state which converts the game field which constitutes that final state into the outcomes which are the basis for emergence.

6.5 WARGAME THEORY

Wargaming can be used to provide organizational senior leadership with a venue in which to consider proposed future operating environments and the complex problems those environments pose at selected knowledge levels. The problem driving this institutional requirement is that of a rapidly evolving operational environment which can outpace the ability to consider implications and imperatives, and enact timely material and non-material solutions. Why use a wargame to address this problem? Because wargaming allows the consideration of events beyond the current moment and is an alternative to computational analysis whose sophistication and complexity are both restrictive and no guarantee of correctness or certainty. Law-like regularities and persistent, predictable dynamics are fleeting occurrences in warfare. This is a characteristic that will only be amplified in the future hybrid operating environment.

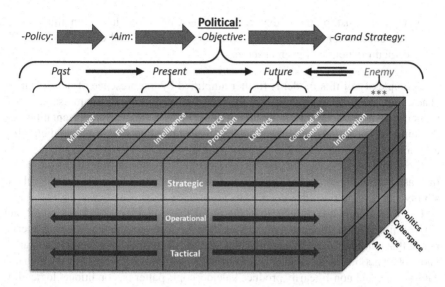

FIGURE 6.3 Model of the complexity of war.

Thus, the purpose of wargaming is to consider and propose the basis for success in a future environment marked by the complexity of multi-domain, information centric, time constrained, distributed, and reflexive warfare by identifying and understanding the prevailing governing factors in such an environment. As such, wargaming is a method of exploration and discovery.

However, war presents us with data that is incomplete, ambiguous, inconsistent, and surprising. Figure 6.3 is a notional representation of this fact. The block represents the relationships that might exist among the levels of war, the warfighting functions, and the domains of conflict as expressed in the 126 areas found in the structure. To reduce the complexity of war to this representation is appealing, but it is not accurate.

As this model moves through time, it is driven by a political process and affected by an operational environment in which the past will dictate the present and the present will become the basis for the future, all influenced by an enemy that will contest, however effectively, in each area. The shading is merely suggestive of the preeminence one side or the other may enjoy in a moment of time and is indicative of an undecided state. Further complication arises when one considers that the layered levels of war are not uniform in motion. The tactical level moves in the space of minutes or hours; the operational level in days or weeks; the strategic level in months or years. Finally, the separate areas in the block operate under a fluid protocol:

1. Areas are not equivalent in value from one time period to the next.
2. As the conflict evolves, the emphasis a side gives to any area may change.
3. The emphasis given to an area may not represent its true value.
4. The positions and distances that exist among the areas are not indicative of their operational relationship.

5. The relationship among the areas is defined by a calculus of simultaneity, sequence, emphasis, value, opposition, convergence, and divergence ... all difficult to define or even observe.

The complexity of this model is further amplified by the presence of non-linearity, chaos, and catastrophe produced by the fact that conflict rarely manifests law-like regularities or predictability in its interactions. Even in this bare representation, the dynamics of the model declare the problem.... absolute knowledge and predictability are foreign to conflict.

If clarity is not a feature of conflict, how does a wargame develop the insights and the basis for intelligent action? How is it possible that any coherence or comprehensibility can emerge from this complexity?

If a wargame is a technique that allows for the exploration of the essential and determining human dimension of conflict, then it is also burdened by the characteristics of that conflict. If law like regularities are rare in conflict; if unpredictable, rapid, and violent actions persist and interact across the levels and domains of war; if chaos, catastrophe, and non-linearity produce kaleidoscopic patterns, conditional logic, illusions, and misdirection not subject to computational prediction and resolution, then it can be expected that a wargame will experience the same dynamics. A wargame will produce a mass of specific and situational information which must be considered in light of a competent assessment methodology in order to obtain value. Thus, the question arises of how to extract information from the dynamic of a wargame that can form the basis for recommendations and further analytical study.

The key to recovering insightful findings in a wargame is to be found in the quality of design. Wargame objectives influence the design and construct of the system of systems which then specifies the "how" and "why" participants engage the game field without actually dictating the substance of what emerges from that engagement. Referring to Figure 6.4, each decision cycle in a wargame is meticulously structured to both accommodate human agility and address purpose. In a cycle, purpose and input fuel a phase of activity marked by the generation of ideas, a controlled period

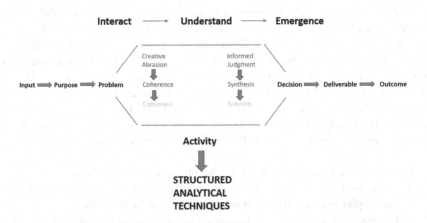

FIGURE 6.4 Decision cycle functionality.

of abrasion as ideas are subjected to collaboration and turbulence in considering the problem, and then a converging toward consensus until a point of synthesis is reached and a decision emerges. This decision is captured in a set of specified deliverables which, combined with germane variables, is subjected to an adjudication methodology to form a cycle output. This output then forms the input for and fuels the next cycle. The design of these cycles is an art in which the problem posed is facilitated by a human and technical environment constructed so that possibilities clash with realities and evaluation produces resolution. Embedded in this art and establishing a substrate for decision, data manipulation, and visualization is the science of computational analysis.

In a wargame application, computational analysis is the extraction of precise information from a complex reality through the use of an organizing mathematical methodology. It relies for its manipulation of data and its precision of result upon the identification and mathematical specification of variables and the quantification of the interactions these variables endure. In this analysis, precise, but not necessarily accurate, conclusions emerge from the connection of a problem susceptible to quantification to a specific method. The attainment of precision and accuracy are laudable and necessary achievements in the study of a weapon's system; it is not a remote possibility in the study of war. As previously stated, there are few actions in war that are law-driven or consist of dynamics capable of isolation from the many ordered layers, effects, and consequences of operational conflict. Therefore, wargaming acknowledges the production of insights that may lack precision or accuracy but can be assessed to possess value in connection with an understanding of the whole that can be augmented and illuminated by the results of a computational process.

The question is: How to associate computational analysis and wargaming to mutual benefit? The answer involves the recognition of the distinct natures of the two approaches and rests upon the effort to forge complementary methods. Wargaming permits judgment to be influenced in a dynamic context by emerging evidence, possibilities, and realities as a precursor to decision. Computational analysis can aid this process through the use of methods that can prepare information before a game, inject points of precision during play which then merge with and act as an informing substrate for decision, and refine outcomes after play to support and enable the transition of defensible insights in support of an informed combat development sequence. Thus, wargaming is not a methodology that can be transformed into pure science by the simple incorporation of more computational analytical tools that push wargaming along an ever more precise spectrum. There is only a particular application of an analytical tool to a specific wargaming requirement that can inform, inject plausibility, and refine outcomes in a manner designed to influence or clarify but not determine game activities and results.

The intent here is to create a relationship between wargaming and computational analysis that benefits both by making both more relevant through subjecting the numerical result to the harsh reality of the human encounter with conflict that is replicated in a wargame. As it is important to recall that just as a wargame makes no assertion as to proof or validation of a solution so numerical precision does not mean you understand the problem or have defined the solution.

This caution highlights both the power (so imposing it seems conclusive) and the limits (unable to comprehend the context it operates in) of computational analysis.

The practice of computational analysis suggests that nothing beyond the numerical result and its immediate context can be verified. Anything beyond is judged to result only from speculation and judgment. Thus, the effort to extend the computational result into a real application with related implications and imperatives is subject to dynamics beyond the comprehension of the process that derived the numerical result in the first place. As asserted, wargaming is not part of some computational analytical spectrum but employs these methods so that wargaming can venture into the kaleidoscope of the unknown with confidence in a method that forces the unknown to surrender what is hidden. It both embraces this engagement of reality and suffers if reduced to a mere vehicle for producing data. Data is foundational, but breakthroughs and realizing hidden coherences in a problem that govern the solution are the province of the human imagination. Thus, sophistication and complexity of method are no guarantee of accuracy and certainty. In fact, it may produce self-delusion in which one accepts the mere appearance of shaping reality and molding outcomes as reality itself. Knowledge does not have to be quantified nor driven by rigid principles in order to be defensible or actionable.

The point here is that, in human affairs, computational analysis is a valuable intellectual construct that can address a sharply defined problem. However, it imperfectly reflects and can significantly distort the reality it is supposed to describe by relying upon the material to the exclusion of the incalculable possible. Wargaming adds the possible.

That said, observation and experience reveal how wargaming professionals are learning to incorporate the methods and results of computational analysis to produce islands of plausibility which can be incorporated into the astonishing flexibility, continuity, and coherence of the human intellect under the stress of conflict. The effort to optimize this relationship entails more than the simple incorporation of MMT into wargame design … this is not new. The struggle is to incorporate the process of computational analysis and the action of the human intellect such that the potential of both are integrated and optimized by using the computational result as a substrate for human decision.

Wargaming is both an art and a science—what craft is not both? As a craft, wargaming encourages creativity, insight, and innovation. Scientific discipline can imply rigidity and produces only the required or the anticipated. Rarely is it brilliant without the addition of imagination. Wargaming is an art because it must not only stimulate imagination but accommodate human intuition and the changes in environment caused by decisions. As an art, the value of a wargame's outcomes is measured by the solution space it delineates and the depth and direction it provides for further study. It is a science because all that is entailed in design must support the interaction of many variables and promote a direction and coalescence of thought focused upon the wargame's problem, purpose, and objectives. As a science, wargaming abides by design standards and methodologies which are the content, enabler, and measure of a successful execution. While computational analytics can both support and benefit from wargaming, it will never replace it and is deficient separated from it.

6.6   NEXT-GENERATION WARGAMING

While the means—wargaming—and the end—to identify the governing factors that pertain to and examine the capabilities and operations required for a success in a complex environment—are understood, the way—how a wargame is to accomplish this—is evolving in the form of a human-centric, technology-enabled style of wargaming art and method. For all the utility and elegance of form expected in wargaming, it is the substance of what transpires and is produced in a wargame that gives meaning and measure to its employment. Thus, the Next-Generation Wargame (NGW) is an advanced, enhanced, and enabled wargaming system of systems.

The NGW is intended to seamlessly represent an evolving operational environment and accommodate the agility, imagination, and disruptive thought of the engaged participants. The NGW is designed to exploit this nexus between human efficiency and wargaming effectiveness. While NGW will employ the latest methods, models, and tools (MMT) to support scenario evolution, decision mechanics, collaborative synthesis, and in-stride adjudication, NGW is not about technology but about facilitating, advancing, and assessing the human processes which drive a wargame. Thus, MMT is an enabler for cognitive and structured analysis and a substrate for human collaboration and decision-making, not a substitute for these things.

There are two pillars, supported by a number of associated principles and technical characteristics, which comprise the NGW. The expression (6.2) below represents the process by which a final game state is obtained by progressing through a series of intermediate game states by employing the principles and pillars of the NGW process. This expression represents both the human cognitive process and the system processes involved in the conduct of a wargame.

$$S_{n+1} = U(s_n, e_n) + D(u_n, c_n) + E(d_n) + R(m_n) \qquad (6.2)$$

The first pillar is the requirement to wargame at the speed of thought. This entails informed deliberations that embrace risk and consider alternatives and thereby produce convergent and divergent thought among the players. The term $U(s_n, e_n)$ in Equation 6.2 represents the *understanding* of the current game situation and is a human social function that involves developing a perception of the game situation that is presented to the players, s_n. The discussion, debate, and analysis that the participants engage in while determining that situation is called creative abrasion and denoted by e_n. The result of using the perception of the current game situation and the expertise of the gamers to understand the game situation is the essence of speed of thought. The key here is the creation of a rapid and fluid environment of creative abrasion in which ideas are generated, interact, are modified, and are either incorporated or are discarded in part or whole as a collaborative result emerges. This dynamic of a rapidly emerging collaborative result will be supported by visualization and planning tools capable of portraying alternatives, consequences, and opportunities.

The second pillar is the necessity to maneuver knowledge: not manage, but maneuver. This entails the deliberate and timely movement of knowledge to a concerned agent for the purpose of achieving a reinforcing outcome. At first glance, this seems little more than the development of effective IT methodologies for the movement, display, and manipulation of ideas at the pace of generation. It is that, but is much more than the narrow search for a technical improvement. This is captured by the term $D(u_n, c_n)$ in Equation 6.2. It considers the understanding resulting from the speed of thought, u_n along with the participants' consideration, c_n, of what is suitable, feasible, and acceptable within the context of the game. Among other advantages, the maneuver of knowledge will be a vital component in the NGW as it is a proposed means of combatting the emerging theory of reflexive warfare. Reflexive warfare (Thomas 2004) involves conveying to a target information and impressions designed to induce a voluntary decision which is contrary to the target's interests. It is a method dependent upon a very sophisticated application of deception theory (Ettinger and Jahiel 2010) and entails the construct of a false image through a calculated distortion of reality. However, it is more than simply the manipulation and reinforcement of perception to maintain an induced belief. It is actually an attempt to transition belief from one state of understanding to another state consisting of a different or even opposing understanding and to manage the resulting decisions by creating a behavior model that can be controlled in the target.

By using an integrated, multi-domain information warfare strategy, an opponent encourages the development of confidence in the target so that he believes that he correctly perceives the true nature and implication of what is being presented. Thus, the target is convinced that he is responding in an appropriate manner. This is an illusion that will be fed, the results of which will be exploited at a later stage. The opponent's intent is to inflict a loss of intellectual and organizational cohesion on the target that impedes a correct response to a situation and so reduces that response to irrelevance. Simply, an opponent intends to establish certain beliefs in the mind of his target, encourage certain actions, and conceal the deleterious outcomes all to the end of inducing a fatal loss of cohesion.

That said, every deception has at least one weakness in that, finally, it's not real. The maneuver of knowledge is expected to take advantage of this by exploiting the impossibility of maintaining the deception across a multi-domain battle space subject to continuous probing, interrogation, reporting, and countermeasures. The maneuver of knowledge will allow situational inferences and ascertainments of ways, means, and ends to be examined from various perspectives and will subject the emerging interpretation of a situation to a comprehensive and collaborative investigation for operational anomalies, all in an enhanced, dynamic, and distributed intellectual environment. The result is expected to be an organization's enhanced ability to fragment and penetrate a reflexive image and to change, formulate, and initiate responses without the loss of cohesion. Failing that, an organization should at least be better able to tolerate a level of approximation and uncertainty if it fails to penetrate a reflexive image. The measure of this outcome will be found in comparing what the opponent intended with what the target actually did. Here, the results of a comprehensive and accurate adjudication process will be essential.

Taken together, speed of thought expressed by $U(s_n, e_n)$ and maneuver of knowledge expressed by $D(u_n, c_n)$ constitute the primarily social elements of wargaming. The remaining terms in Equation 6.2 $E(d_n)$ and $R(m_n)$ refer to execution of the decision determined during maneuver of knowledge and the resolution of the move, i.e., adjudication, respectively. These last two terms are primarily physical processes, often implemented by machines or performed by formal processes. When wargaming is viewed as a combination of human social processes as well as technical physical processes, wargaming quickly rises to the bar of an interactively complex system of systems with ample opportunity for emergence.

Wargaming rests upon too many interacting variables and fluid situations to permit the assembly of a final emergent behavior without intermediate emergent compositions. Emergence in wargaming is a combination of sub-emergences whose values, emphases, and relationships will change with the evolution of game states. Narrative emergence is the "story" the participants build in a game as it progresses.[1] It is vital as a means of establishing connection, coherence, and purpose as the game states evolve and the scenario develops. Dynamic emergence is the change in the complex operating environment under the duress of decisions and interactions which cause permutations to influence the value, emphasis, and relationships among selected areas of action. Intermediate emergence is the local production of adjudication results that occur to influence the formulation of a new game state. As this process is continuous in the life of a game, a cumulative emergence will be generated as the sum of and final expression of the combination of narrative, dynamic, and intermediate emergence in the definition of and assessment of the final game state.

The cumulative emergence of a wargame contains the summation of the wargame's emergent behavior which resides in the game's final state and which is extracted with and given substance to by cognitive and computational evaluation processes. The final state of a wargame is represented by the expression presented in Equation 6.3 and below. Here the actions of the speed of thought upon cognitive calculation and consensus and the maneuver of knowledge upon informed judgment and decision to produce emergence are associated so that, combined with adjudication and a resulting evaluation, the current state of a wargame is defined. The sum of these continued and repeated actions eventually produces a final game state such that: ε_c is the cumulative final emergence; S_N is the final game state; (e_N, e_D, e_I) are the narrative, dynamic, and intermediate emergences; and E is the final cognitive and computational analysis, evaluation, and assessment procedure.

$$\varepsilon_c = S_N(e_N, e_D, e_I) \cdot E \qquad (6.3)$$

6.7 ADJUDICATION

The driving mechanic of a wargame is to force human decision-making, synthesize the cumulative effects of many decisions, and then represent the resulting outcomes and consequences so that subsequent decisions are required. A comprehensive portrayal of diverse and contending wargame consequences does not lucidly arise as an ordered whole from even a deliberate assembly of arrangements, decisions, intents, and resources. In order to arrive at that ordered whole, which then acts as a substrate for continued game play and subsequent decisions, a wargame must provide for an appropriate adjudication process. This not only brings a wargame by stages to culmination, it is also the basis for producing the final state variable in Equation 6.1.

If wargaming is a narrative containing a dilemma, then adjudication is an act of recurring resolution of that dilemma which advances the narrative. Adjudication entails the gathering of diverse, opposing, divergent, and complementary decisions

---

[1] Thank you to our wargame colleague, Mr. Mark Gelston, for reminding us of the importance of narrative in a wargame.

and actions; subjecting them to a process which considers timing, intent, movement, and weight; and then presenting a coherent picture of results which drive the subsequent conduct of the wargame. Adjudication is nothing less than the means by which the uncertainty of an initial state is transformed into a coherent understanding of a subsequent state which is then subject to a continued trajectory based upon ensuing player decisions and interactions. Of all the requirements for a successful wargame design, the selection of the proper adjudication method is the most crucial.

A seminal white paper by Wiggins (Wiggins 2014) lists four types of wargaming adjudication methodologies:

1. Rigid Adjudication relies on strict rules and systems from which products emerge that provide the results of player interaction.
2. Free Adjudication is an interpretation of player interaction in light of experience, expertise, and history.
3. Semi-Rigid Adjudication is a combination of the previous two which mixes system results with expert interpretation to arrive at interaction results.
4. Open Adjudication relies upon the players themselves to consider the governing factors of a situation and produce an agreed upon resolution.

While these are recognized types of and approaches to adjudication, they are individually not sufficient to satisfy the advanced demands of NGW. This requires the formulation of a method called "in-stride" adjudication. In-stride adjudication is the synthesis and evolution of all these methodologies in a hybrid form which permits the combined application of human judgment with MMT techniques driven by the specific requirements of a continuous, evolving, and dynamic wargame operating environment. In-stride adjudication is the resolution of the initial state-subsequent state problem based upon continuous, asynchronous decision cycles instead of a simultaneous move mechanism (e.g., football) or an application of a sequential move mechanism (e.g., chess). Continuous implies that no discrete move limits the number or extent of actions possible; asynchronous implies that a team advances its action at will until an opposing action denies a purpose or seizes the initiative thus forcing a reaction. In a word, in-stride adjudication resolves player interaction as it occurs and as many times as it occurs using a fluid combination of rigid, free, semi-rigid, and open adjudication methodologies supported by MMT.

Understanding this, the very act of in-stride adjudication becomes a means for considering the principles, methodologies, and implications of reflexive warfare. The measure of effectiveness of a reflexive warfare operation is the decision it induces. In a word, does the target do what the opponent intended? Reflexive warfare is not simply the action of establishing and reinforcing a false image, but, instead, it is a process which progresses incrementally and manages the evolution and migration of this image to induce a calculated end. Thus, in-stride adjudication's continuous and asynchronous resolution of a dynamic situation will provide a control team with the ability to know ground truth, understand, let's say, Red's reflexive strategy, observe Blue's developing interpretation, and monitor Blue's ensuing decision and its result. The effectiveness of the reflexive action and the target's susceptibility for or resistance to a loss of cohesion can thus be gauged by measuring the difference between what Red intended and what Blue did. This will permit a wargame to identify and explore the principles and methodologies of and the effective protocols to counter, reverse, employ, or exploit a reflexive warfare operation.

In this way, in-stride adjudication supports and facilitates the pillars of the speed of thought and maneuver of knowledge through a gaming environment which employs distributed and interactive visualization, collaboration, and information merging and manipulation capability.

This will permit the in-stride resolution of interaction and the customization of the results relayed to each team. These continuous, asynchronous decision cycles involving blended action-reaction-counter-action phases experienced by the players will encourage the making of estimates, assumptions, the calculation of risk, and action under the simulated pressure of operational tempo. Governing of this operational tempo is exercised through MMT designed to influence the pace of a wargame through the quality, quantity, completeness, and availability of information and resources provided. Game time can be actively modulated according to the phase, intensity, and duration of an operation. This method permits an adjudication team to guide the flow of a game through the collection of information, the assessment of probabilities of occurrence and success, and the nonintrusive observation of evolving work in player cells. Subjecting player submissions to the adjudication process both resolves an action and permits control to customize its response in light of deception, ISR, logistics, fires, technical and communication denial, cyber-attacks, and combat results.

In-stride adjudication is an attempt to generate measurable momentum in game play. It posits continuous, asynchronous play at the speed of thought producing the maneuver of knowledge in order to promote intellectual exchange, creative abrasion, and collaboration among players and the teams they constitute. In-stride adjudication sustains this process by forging a union among the different methodologies of adjudication so as to act as a means of measuring how player participation changes the state of the game and how teams reconcile the encounter of situation and task. This synthesis of methodology, action, contributory thought, alternatives, and resolution aspired to in a collaborative gaming environment will permit the study and assessment of the anatomy and the autopsy of game decisions …anatomy being the elements of purpose, method, expectations, and desired end state that influence and comprise a decision; autopsy being the resulting outcomes, consequences, and responses to the evolving situation that define the force of a decision. NGW then becomes a point of synthesis combining the informed participant, the collaborative environment, in-stride adjudication, and an MMT substrate into a human-centric, technology-enabled gaming architecture and design configuration that offers both a means for stimulating emergent behavior and a potential evolutionary path for the art and science of wargaming. In a word, adjudication, but in particular the advanced technique of in-stride adjudication, is the forcing function which generates and gives form to emergent behavior.

6.8 THE FUTURE: AN EXAMPLE OF THE APPLICATION OF EMERGENCE

He who is everywhere is nowhere.

(Seneca the Younger)

Frederick the Great, a student of Roman literature, may well have adopted his famous "He who defends all defends nothing" from Seneca. If so, we may take the liberty to say that "He who prepares for all contingencies prepares for none." And yet, prepare one must. This implies a sense of the future and a reasonable appreciation of the competitive environment that will exist and the advantage that must be acquired for success. Thus, emergence of the future requires the identification of ends, the selection of ways, and the securing of the means based upon the projected conditions for success in a future operating environment. The key is to "learn fast enough to sustain a competitive advantage" (Mason 2014).

What then is the future? Avoiding the deep scientific and philosophical implications of that loaded question, the answer to this question for the wargamer is to be found in understanding the past. History can be understood as the marriage of events and their causes which merge and emerge to produce the trends that comprise the present … the here and now. Trends are "concrete forces that … have high impact on the strategic environment and are highly predictable over the planning horizon" (Connections US Wargaming Conference 2019 2019). Examples of trends, which are emergent behaviors in and of themselves, are demography, technical proliferation, and information as a weapon. The present is the synthesis of these past trends and their consequences which have been modified by the friction of an historical context (political, economic, military, cultural, etc.) that both limit and shape the extent of what goes forward. In recognizing this dynamic, the future is understood to be a continuation of this process marked by the sensitivity, variation, and uncertainty which are characteristics of a complex process—a process determined by the interaction of time, historical context, available material, operative ideas, and human action that makes for a capricious path forward. This complex process means that it is easy enough to identify the current general trends and an approximation of the principles that drive them, but difficult to gauge the emerging trajectory in time. This is so because, as obvious as general tends can be, they are connected by interlocking dynamics that are immersed in change, disruption, and hidden details so as to prevent precise calculation.

Thus, the future represents an evolving contingency marked by the extremes of uncertainty, sensitivity, and variation with confluences at critical junctures which can exaggerate or suppress estimated deviations and deflect expected trajectories. The attempt to predict the future is marked by certain miscalculation, accident, and frustration. The secret to considering the future is to forsake the standard of perfection and seek plausibility based upon a defined spectrum of emerging outcomes. In this endeavor, wargaming can be of assistance.

As a wargamer considers applying his art to the problem of the future, he quickly realizes that the problem is not "can I wargame the future?" but "how do I wargame the future and to what benefit?" A wargame that considers the future can be designed with any competitive environment desired and with a proposed problem to be addressed. You want to examine a future inter-planetary war driven by conflict on Mars? Simply construct the environment and state the problem. In short order, a scenario and supporting design can be assembled and the study can commence. SPI's "BattleFleet Mars" from 1977 is an instructive example.

However, the problem here is obvious. Because the game is separated from all but the most universal of trends and is finally based upon construct, supposition, and speculation, the results of the game are of dubious benefit to an analysis of future environments and associated emerging governing factors and requirements. This is so despite the fun one may have had commanding a battle fleet of space ships. Again, the problem is not "can" I wargame the future but "how and for what benefit."

At this point, a synthesis of skills must be proposed. Separated, the futurist and the wargamer are devoid of the necessary skills to examine the future and extract both insights and direction. The futurist is one who possesses the skill and foresight methodologies necessary to both construct and evaluate an ever receding set of plausible alternative futures based upon general trends and associated assumptions that exhibit both a plausibility and a rate and level of change. The wargamer can then take these proposed futures; extract problem, hypothesis (if … then), and statements of objective based upon the projected operational environments; and design the vehicle necessary to consider this dynamic context and the associated proposed responses required to operate successfully in those environments. In a word, the wargamer can employ the work of the futurist in a construct that permits the examination of an emerging future and the responses to that future given a definition of what constitutes success. This is "how" to wargame the future.

Complexities will abound in this process because of the need to coordinate intelligence estimates; political, cultural, economic, and demographic trends; government policy; scientific and technical progress; and shifting relations in power. Thus, a futurist and a wargamer must experience a fusion of both purpose and talent in projecting the future environment and designing a wargame to consider it. The wargame will become a kind of solvent in which disparate elements, estimates, and speculations about level, kind, rate of change, and interaction among variables will have to cooperatively co-exist. A good futures wargame will require many coordinated qualifications that address assumptions, account for factors, and define contingencies. The futurist and wargamer will have to live in each other's fields, and this confluence should have the added benefit of strengthening the understanding and reinforcing the cooperation between these two fields of practice.

Like the wargamer, the futurist relies upon a stable process for considering the future but is confronted with a spectrum of ways in which that process can be employed in the service of trends and objectives. An initial consideration is to distinguish between forecasting the future, which entails following a collection of trends with estimated deviations and arriving at a point, with the development of alternative futures, which entails developing a collection of alternative strategic environments that bound a region of plausible outcomes each characterized by an overarching descriptive title that is developed in detail in the accompanying scenario document. Figure 6.5 provides an example of the possible alternative strategic environments that might emerge in a hegemonic world in which economic or military power is the governing factor (the scope) and the extremes (the scale—dominant or subservient) of the US position in such a world are the matters to be considered.

Peter Bishop (Bishop and Hines 2012) poses a series of questions and answers that act as a guide to the selection of a methodology for thinking about the future. Without exploring the deep value contained in this approach, several highlights will serve to establish some parameters. A defining idea is that the best way to think of the future is not to expect to be right, but to avoid being surprised. This strengthens

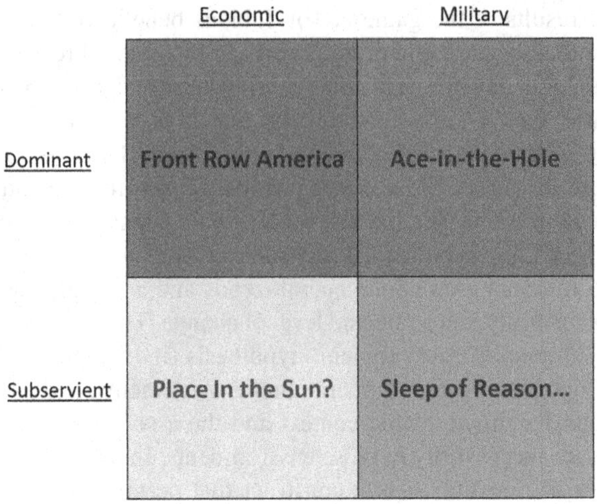

Economic   Military

	Economic	Military
Dominant	Front Row America	Ace-in-the-Hole
Subservient	Place In the Sun?	Sleep of Reason...

FIGURE 6.5 An example of U.S. alternative strategic environments in a hegemonic world.

the desirability of considering alternative futures rather than trying to achieve a precise forecast. Thus, one is led to consider the value of probable, plausible, and preferable futures. He states that a probable future is one in which surprises don't greatly affect the trajectory of time; a plausible future incorporates the consideration of and emphasis upon selected variables which expand the range of outcomes; a preferable future is the definition of an optimum future state in which advantage and benefit are maximized. The latter represents an interesting condition in which context, material, and ideas coalesce to define a consensus goal which informs concerted action. This suggests a systematic way in which action today determines outcomes tomorrow. In a word, while it is not possible to obtain an accurate prediction of the future, it is certainly possible to deliberately influence the future and achieve what is preferable to a degree. The point here is not that this idea is new. After all planning is an ancient art. The point is that by systematically considering alternative futures, assumptions can be refined, implications can be anticipated, and a design and priority of work can be arranged and coordinated such that a preferable future is more likely to result. Peter Bishop concludes by emphasizing the importance of the result of this process: the establishment of a general direction for fundamental change.

From a wargamer's perspective, how might the identification of this advantageous fundamental change be achieved and the path to attainment be established? One course of action, which can be considered in the fusion of the futurist and wargamer's methodology, involves the identification of the future operating environment and the advantages required for success (futurist methodology) combined with the ability to formulate, examine, and assess the direction to be taken, the advantages to be gained, and the exploitation to be achieved (wargamer's methodology). A wargame will permit a close inspection of the core principles needed to influence the environment and obtain advantage. These core principles are the governing factors which drive the dynamics of the future environment. Governing factors delineate the operational functions and trends that must be capitalized upon to achieve success. The definition of success and its associated metrics provide the basis for assessing the misdirections and distractions that can be induced by the kaleidoscope of future events, patterns, and tendencies exhibited in the competitive operating environment. Knowledge of this dynamic, if not understanding of the mechanism, is what can convert governing factors into operational acts of persuasion or imposition that generate conditions for exploitation. Thus, while the core principles specify the dynamics of an environment, the governing factors are those elements that can be manipulated, combined, and arranged to impose, persuade, induce, or deceive an adversary into a commitment that can be exploited in the service of a future to be obtained. This places an adversary in a counterintuitive position where direct actions arrayed against him can organize an environment in unpredictable ways, cause risks to be exaggerated or obscured, turn strength into weakness, and cause counter action to be miscalculated and so generate transitory moments in which counter strikes can prevail. This is the benefit derived from a wargame that considers the future.

Thus, a productive fusion emerges. While the futurist identifies the trends that will produce an operating environment, the wargamer sets this environment in motion driven by appropriate dynamics to identify the governing factors whose proper arrangement and utilization become the key to understanding and anticipating the conditions required, the preparation necessary, and the consequences expected in gaining advantage and success in the future operating environment. This is the "how" and the "benefit" fused. This process is made more powerful when several alternative futures are examined in an integrated series of wargames and governing factors common to a subset of or to the whole of the alternatives are identified. Figure 6.6 is a preliminary illustration of the proposed fusion between the art and science of the futurist and the wargamer.

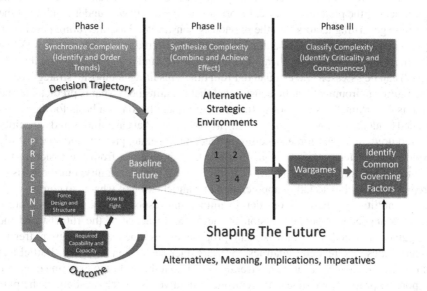

FIGURE 6.6   The fusion of the art and science of the futurist and the wargamer.

The phases of this process establish a construct in which identified trends and attendant uncertainties are converted into a classification of prioritized criticalities and resulting emerging environmental, operational, and functional consequences.

Phase I (Synchronize Complexity): From the perspective of the present, categories of trends and uncertainties are identified and ordered by the power and ability to drive change. This becomes the basis for the development of a decision trajectory in which the futurist selects among weighted alternatives, evaluation of effects, and forward direction to define an emerging baseline future.

Phase II (Synthesize Complexity): Having established a baseline future, trends and drivers are now combined, organized, and arranged into constructs which produce alternate strategic environments (e.g., The Hegemonic World, The Political World, The Technological World, The Resource Constrained World, etc.). These scenarios are developed using a planning process such as proposed by Jay Ogilvy (Ogilvy 2015) and that creates environments which should be constituted to cover a spectrum of possible futures and may focus on probable, plausible, or preferable futures which permit an examination of alternative possible strategic environments that are created by emphasizing different combinations of trends found in the baseline future.

Phase III (Classify Complexity): With the completion of an integrated series of wargames that examine the selected alternative strategic environments, assessments will produce the identification and isolation of the resulting governing factors which represent the key operational functions and trends in that particular environment. With governing factors in hand, the classification of the criticality and operational consequences in that environment can be identified. Finally, the identification of those governing factors that are common to all or a set of the strategic environments examined can be assembled.

With these three phases accomplished, a feedback mechanism is employed allowing for action to be initiated that can shape the strategic environment and influence the attainment of the preferable future. Incorporating the newly won understanding of the common governing factors into the previously constructed baseline future permits an approximation of what must be done to obtain a preferred result. Alternatives (What if?), meaning (So what?), implications (What's next?), and imperatives (What must be done?) can be assessed and evaluated. From this, a resulting composite strategic baseline future environment can be constructed that contains both what is predictable and what is uncertain. This composite future environment becomes a basis for the action needed to identify "how to fight" in that environment, what capabilities and capacities are required, and what force structure is needed to operate, prevail, and successfully exploit that strategic environment. In a word, what emerges from the fusion of the science of the futurist with the art of the wargamer is not certainty, but the basis of preparation which is found in appreciation and anticipation of what is to come.

The ability to generate, consider, examine, and respond to the emerging future world is dependent upon the fusion of the art and science of the futurist and the wargamer. Isolated, the futurist can propose environments based upon trends, uncertainties, and related dynamics; the wargamer can design and play "BattleFleet Mars." But, the study of future environments without the ability to examine dynamic responses, propose outcomes, and consider initiatives is merely academic; the play of a game without an environment grounded in plausibility, oriented against a real problem, adopting an assayable hypothesis, and considering specified objectives is merely fun. Separated, the future process and the wargame lack animation; fused, the combination is one of power and utility. The futures wargame will harness the ability to examine the science of the necessary with the art of the possible in a future operating environment. Most importantly, the futures wargame could be instrumental in dispelling the illusion of certainty that sometimes emerges in a bureaucratic rationality burdened by limited understanding and capacity, constrained information and vision, and driven by external urgencies that compress both time and deliberation.

The more an organization can represent a future competitive environment and consider the dynamics of that space, the more likely that organization is to get the emerging world "roughly right" instead of "precisely wrong." It is much more profitable to shape a response to the former rather than the latter. Seneca the Younger would be pleased.

6.9   WHERE ART AND SCIENCE MEET

Wargaming presents a unique system of systems challenge. As we have shown, not only are the mechanisms of wargaming multiple and diverse, but the very nature of the problems a wargame is designed to address have emergence as a principal characteristic. As we have shown, the nature of war is highly dynamic, and therefore, a wargame itself must be a complex dynamic system. It is comprised of highly autonomous constituents, e.g., the human players in the wargame, and analytical non-human components in the MMT that facilitates the wargame. This heterogeneous mix of humans and machines forms a very complex system of systems. The ability to recognize the complexity in the system that enables game play is essential not only to game design, but also in establishing confidence that the results of the wargame are a reasonable representation of real-world conflict. Fortunately, progress is being made into computational tools that reveal how entities, through their behaviors, can interact to produce aggregate effects. These resulting behaviors are emergent in that they result in uniquely identifiable behaviors that none of the contributing entities would have produced in and of themselves. Providing the human constituents of a wargame with a technological means to rapidly visualize, understand, and explore the emergent results of their choices has very profound importance to wargame designers, players, and analysts. One such tool that has shown promise in exploring how entity behaviors can interact to produce new emergent behaviors is the MP-Firebird tool developed by the Naval Postgraduate School (NPS n.d.). Firebird implements the Monterey Phoenix language that allows entities to be defined by their behaviors and then reasons about how these behaviors can interact. Firebird allows constraints on the behaviors to be defined and so the collective behavior of the entities can be examined, explored, and controlled. Originally developed to support software and hardware architecture development, Firebird is finding many uses in complex systems that include both human decision-making components and well-defined system processes (Giammarco, Carlson and Blackburn, Verification and Validation (V and V) of System Behavior Specifications 2018). MP-Firebird is adept at the modeling and simulation of system of systems that combine multiple diverse domains—and this is what makes it interesting for use in wargaming.

Without delving into the specifics of the Monterey Phoenix language, suffice it to say that MP-Firebird (we will just refer to it as MP from here on) allows us to focus on the entities that make our scenario and the behaviors they exhibit—or at least what we think they will exhibit. Where the idea of emergence arises is that not only can we express entity behaviors in MP, but we can also see how those behaviors interact, producing a new collective behavior. If the collective behavior is what we were intending, then well and good. But if the collective behavior is something we did not anticipate, then we probably should reconsider our entities and behaviors. When MP runs our model, it generates "traces" that represent the various sequences of combinations the behaviors can produce. Although it is easy to appreciate that it doesn't take many interacting behaviors to have our model fall to the curse of dimensionality, i.e., the combinatorial explosion that could be produced, MP relies on something called the "small scope hypothesis," a heuristic method that draws on formal methods but exploits automation to find "flaws" in the model as early as possible. The small scope hypothesis argues that most unintended events (flaws) can be discovered with a relatively small number of examples (Jackson and Damon 1996; Oetsch, et al. 2012; Andoni, et al. 2003; Giammarco and Troncale, Modeling Isomorphic Systems Processes Using Monterey Phoenix 2018). MP not only provides a formal language for expressing such models, but also presents a graphical representation of the traces the model can produce. This visual reorientation enables the modeler to quickly see the consequences of the interactions. The modeler can then impose constraints on the entities to eliminate unwanted behaviors and assure desired behaviors. In a sense, MP-Firebird allows the systems modeler the ability to explore emergent behaviors and play with the balance of variety and constraint to see the consequence, that is what emerges, from the system.

Let's consider how MP-Firebird can be used in a wargaming context. Consider a tactical situation that is just one scenario in a wargame. In this scenario, depicted in Figure 6.7, a forward observer surveys two enemy targets. One is a fixed enemy

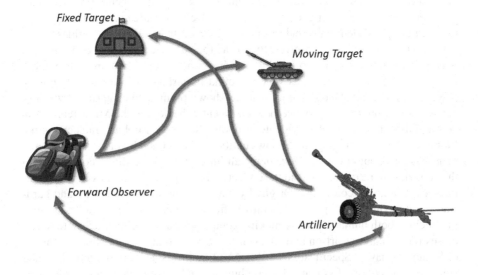

FIGURE 6.7 A tactical situation in a wargame.

position, and the other is a moving enemy target. The forward observer has the job of identifying a target, informing an artillery battery about the type of target and its location, and then observing the results of the artillery's action on the target. MP facilitates our understanding of the tactical situation through formal expression along with an accompanying visualization of that understanding. MP will also "run" the model of our tactical situation and reveal the possible scenarios this situation may produce.

We can start our MP exploration of this scenario by using the MP language to express our entities and their behavior. A model of the scenario is shown in Figure 6.8 MP uses the term SCHEMA to label a model, and this one is called Artillery_Fire_Battle_Damage.

Here we have created entities called Forward_Observer, Artillery, and Target. The keyword ROOT identifies each entity in our model. (Technically, MP refers to these as "root events" so if we were being pedantic we would consider Forward_Observer as the forward observer event which in turn is comprised of other events.) Forward_Observer has the behaviors of *observes*, *calls_for_fire*, *assesses*, and *adjusts*. Similarly, Artillery has the behaviors of *register* (establishing an aimpoint based on information from the Fordward_Observer), *clear_mission* (completing all checks of the area so that firing can be done safely and appropriately), and *fire* (deploying rounds on target). Rather than create two distinct entities for the targets, here we see Target as an entity. Target can be typed as either *Moving* or *Fixed*, and it can be either *Friendly* or *Unfriendly*. Modeling Target in this manner allows for other potential targets other than the two in the scenario description, e.g., a target might be a Fixed Friendly entity (probably a target we wouldn't want to shoot!).

```
SCHEMA Artillery_Fire_Battle_Damage

/* Actor Behaviors */

ROOT Forward_Observer:  (* observes *)

                        (
                          calls_for_fire
                          assesses
                          adjusts
                        );

ROOT Artillery: ( register
                  clear_mission
                  fire
                );

ROOT Target: { (Moving | Fixed), (Friendly | Unfriendly)
               ( cover_effective | cover_not_effective ) };


       cover_not_effective: ( Missed | Damaged | Destroyed );
       cover_effective: ( Missed );
```

FIGURE 6.8 An MP model of a scenario.

MP lets us run our models at our choice of scope. Scope in MP refers to the number of loop iterations we will allow to occur when the model runs. This is an important control to deal with the aforementioned "curse of dimensionality." For example, the first behavior of the Forward_Observer is *observe*, but we have included the symbol "*" on both sides of *observes*. This tells MP that *observes* is a repeated behavior to be done zero or more times. This seems to make sense given that our forward observer should always be observing, but might not sometimes. If we run our model at scope 1, MP produces 32 "event traces", i.e., 32 distinct possible scenarios involving our entities. Figure 6.9 shows three of these possible scenarios generated by MP.

We can see that the Forward Observer calls for fire and does some assessment and adjusts. The Artillery registers, clears the mission, and fires. We can also see that the Target is moving, that it is a friendly target but it was covered and was missed in the top trace and destroyed in the middle and bottom traces. Clearly, we should not be firing on friendly targets. From the top two traces, it appears that calls for fire are being issued without observation. Also, it is easy to see that no coordination is going on between the Forward Observer and the Artillery. What we need to do is to impose some constraints on our behaviors in this model. This is observable at scope one. Figure 6.10 shows a result running at scope 5 (which produces 96 traces) where we can see the forward observer repeatedly observing, an unfriendly target is destroyed but again there is no coordination in the behaviors.

MP supports constraints on behaviors through the COORDINATE keyword. If we add the constraint as shown in Figure 6.11,

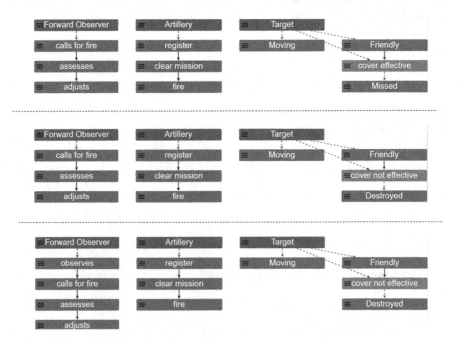

FIGURE 6.9 Scope 1 (32 traces)—Top trace fired on friendly and missed, and bottom trace fired on friendly and destroyed.

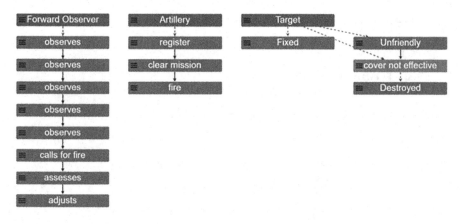

FIGURE 6.10 Scope 5 (96 Traces).

```
COORDINATE  $a: ( calls_for_fire ),
            $b: ( Unfriendly )
   DO ADD $b PRECEDES $a; OD;
```

FIGURE 6.11 Call for fire limited to unfriendly target.

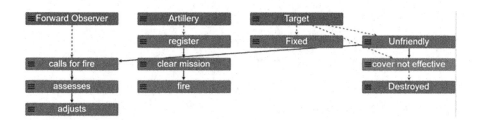

FIGURE 6.12 Scope 1 (16 Traces)—Only unfriendly targets are fired on.

We are telling our model that the behavior *calls_for_fire* can only occur if the Target is *Unfriendly*. Adding this constraint reduces the number of traces at scope one to sixteen. Figure 6.12 shows an example. But we can still see that there are unwanted behaviors that require constraints. Here we can see that the forward observer is still not performing the observe behavior, and there is still no coordination with the artillery.

The correct sequence of activities for the entities in this model would be for the forward observer to observe until an unfriendly is seen, then call for fire to the artillery. The artillery would then register the target, clear the mission, then fire. Finally, the forward observer would assess the target and adjust if it was missed or return to observing if was destroyed or killed. The following model and set of constraints in Figure 6.13 achieve this correct sequence. Notice that we did make some changes to the original model by adding the *mission_accomplished* event. By making *mission_accomplished* explicit, we are able to verify the scenario where a target is damaged or destroyed.

```
SCHEMA Artillery_Fire_Battle_Damage

/* Actor Behaviors */

ROOT Forward_Observer:
                        (
                          observes
                          calls_for_fire
                          assesses
                          (adjusts | mission_accomplished )

                        );

ROOT Artillery: ( register
                  clear_mission
                  fire
                );

ROOT Target: { (Moving | Fixed), (Friendly | Unfriendly)
               ( cover_effective | cover_not_effective ) };


     cover_not_effective: ( Damaged | Destroyed );
     cover_effective: ( Missed );


/* Constraints */

COORDINATE  $a: ( mission_accomplished ),
            $b: ( Damaged | Destroyed )
   DO ADD $b PRECEDES $a; OD;

COORDINATE  $a: ( Unfriendly ),
            $b: ( fire )
   DO ADD $b PRECEDES $a; OD;

COORDINATE  $a: ( Artillery   ),
            $b: ( calls_for_fire  )
   DO ADD $b PRECEDES $a; OD;

COORDINATE  $a: ( Target   ),
            $b: ( observes  )
   DO ADD $b PRECEDES $a; OD;
```

FIGURE 6.13 Scope 1 (6 Traces)—More complete constraints. Only unfriendly targets are fired on, adjustment only if missed.

When this model is run it produces six traces. Figure 6.14 shows two traces. The top trace is the situation where the target is effectively covered, missed by the artillery, and the forward observer adjusts. The bottom trace shows the situation where the target is destroyed, signifying a completed mission.

It is important to note that this example is not necessarily complete; there are many different behaviors that could be considered and examined in MP. It's presented to show how a tool like MP, originally created for model testing in software development, is well suited to the examination of behaviors in system of systems. In a wargame, a tool such as MP can be used in almost any aspect of the wargaming process. It could

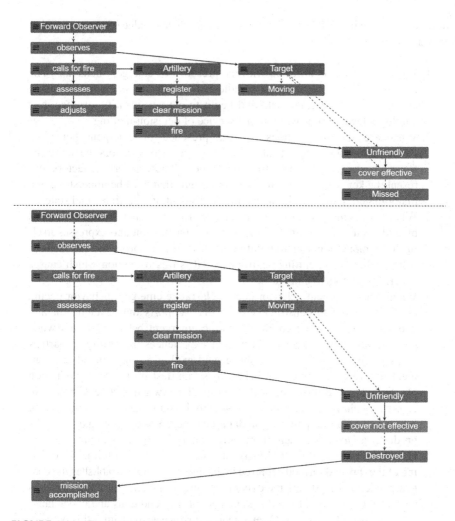

FIGURE 6.14 Two traces from the constrained model.

be used by wargame designers to identify key features of a wargame that they want to emphasize to participants. It could be used by participants to explore various possibilities that they want to control. It could be used by wargame adjudicators to assess likely outcomes and compare those to participants' choices in the wargame. In all cases, MP provides people the ability to better detect and understand the implications that might emerge when the different behaviors of their systems interact.

6.10 CONCLUSIONS AND WAY AHEAD

A wargame is an interactively complex process in which the involvement of the human intellect produces an emergent behavior that is unpredictable and rich in content. Acknowledging the existence and the value of emergence in wargaming, six questions emerge in considering the nature, operation, and value of emergent behavior in wargaming:

1. What actually is manifested with the presence of emergent behavior? The presence of emergent behavior is indicated when participants in a wargame engage a dynamic game field and begin to extract, identify, arrange, and employ relevant information in the service of decision-making.
2. What are the implications for the presence of emergent behavior? Participants obtain an understanding of the governing factors that drive the functions and trends in a hostile environment. These governing factors represent the key operational functions and trends that must be understood and manipulated in order to influence the attainment of a desirable end state.
3. Where does emergent behavior take place? Emergent behavior is conceived by and occurs in the mind of the wargame participant and expresses itself in the computational and cognitive analysis that resolves the interactions generated in the operating environment. It is this resolution which causes governing factors to emerge.
4. When does emergent behavior occur? Narrative emergent behavior occurs as the participants construct a story that supports the development and understanding of the scenario. Dynamic emergent behavior occurs when the wargame methodology detects and experiences the ability of participants to successfully modify, shape, and exploit an operational environment. Intermediate emergent behavior is manifest in the local production of adjudication results that form the basis of a new game state. Cumulative emergent behavior occurs upon game completion when the combination of narrative, dynamic, and intermediate emergence with other game results produces a final assessment from which emerge wargame outcomes.
5. How does it manifest itself? Emergent behavior manifests itself in the ordering of the flux and turmoil captured in a game field which establishes the context needed to investigate the dynamic nature of a game's problem and then enables the extraction from the game field of the conditions and understandings necessary for success. Often enough, surprise accompanies this process.
6. Given the presence of emergent behavior is considered to have negative effects what is done to control it OR given the presence of emergent behavior is considered to have positive effects what is done to capitalize on it? Wargaming does not suffer a distinction between positive and negative emergent effects. Emergent behavior is elicited in response to the chain of a wargame's construct—problem, objectives, design, participation, narrative, dynamic, intermediate, and cumulative emergent findings, and assessed outcomes. What emerges from a wargame is valued for its ability to modify or extend the understanding of the given problem. Thus, the emergent behavior itself is neither controlled nor capitalized upon; it is assessed and incorporated into the understanding of the problem and into the refinement of the shape of the emergent governing factors that constitute the substance which drives the problem. It is the resulting definition of the governing factors that are controlled or capitalized upon by the relevant organizations.

The consequences derived from the coalescence of these six elements is not only the detection and certification of the existence of an operating emergent behavior but also the successful action of its transformative properties in a hostile environment. Thus, the principle of emergence can be substantiated in the operation by which participants in a wargame engage a dynamic reality and extract, identify, arrange, and employ the governing factors necessary for successful operation. If surprise characterizes and the unexpected accompanies this process, one can be fairly certain that emergent behavior has transpired.

The problem in generating emergent behavior through the adoption of advanced and comprehensive wargaming methodologies and supporting technologies that will provide marked improvements in design, execution, and assessment is twofold: "How do we measure if the wargame is actually performing as expected and is the wargame delivering emergent behavior outcomes that can be justified?" The previously mentioned principles of wargaming at the speed of thought and the maneuver of knowledge are the key to realizing the potential of this advanced wargaming technology nested in the collaborative gaming environment. The difficulty lies in understanding the exact composition of the two principles in action and in developing the metrics necessary to determine when and to what extent speed of thought and maneuver of knowledge are happening in a wargame and how this influences what emerges as outcomes. Alignment of objectives, design, execution, and outcomes is a measure of efficiency; demonstration of quality of outcomes is a measure of value. The solution is to be found in developing metrics that employ both graph theory and cognitive science to quantify these wargame actions in terms of efficiency and to realize that value is a quality evaluated in post-game study, experiment, and exercise. The benefit of these metrics is the ability to certify the efficiency of a wargame's execution and so to measure the resolution, depth, and fidelity of emerging outcomes in order to justify confidence in the effort to examine the value of outcomes in subsequent work. Thus, this development and certification of a metrics protocol will allow both the efficiency of a game to be measured while also providing a basis for the justification and substantiation of the value of the wargame's emerging outcomes. An added benefit is the identification of best practices, useful MMT, and improved methodologies. As industry's integration and development of an advanced wargaming architecture matures, the development of a metrics' theory and prototype should be pursued and incorporated into the architecture so as to perfect the metrics' application and utility. In this way emergent behavior in a wargame can be identified, evaluated, and its value pursued in subsequent study.

It is through the patient assessment and consideration of what emerges in the execution of a wargame that Frederick the Great's insistence upon the value of reflection is both realized and beneficially applied.

BIBLIOGRAPHY

Ablowitz, Reuben. 1939. "The Theory of Emergence." *Philosophy of Science* 6 (1): 1–16.
Andoni, Alexandr, Dumitru Daniliuc, Sarfraz Khurshid, and Darko Marinov. 2003. "Evaluating the 'Small Scope Hypothesis'." In *POPL '02: Proceedings of the 29th ACM Symposium on the Principles of Programming Languages*, Vol. 2.

Bishop, Peter C., and Andy Hines. 2012. *Teaching about the Future*. New York: Macmillan.

Connections US Wargaming Conference 2019. 2019. *Wargaming the Far Future*. Working Group Final Report, Carlisle PA: US Army War College.

Ettinger, David, and Philippe Jehiel. 2010. "A Theory of Deception." *American Economic Journal: Microeconomics* 2 (1): 1–20.

Giammarco, Kristin, Ron Carlson, and Mark Blackburn. 2018. *Verification and Validation (V and V) of System Behavior Specifications*. Technical Report, 01 Nov 2016, 31 Oct 2018, Monterey CA: Naval Postgraduate School. https://apps.dtic.mil/sti/pdfs/AD1063328.pdf.

Giammarco, Kristin, and Len Troncale. 2018. "Modeling Isomorphic Systems Processes Using Monterey Phoenix." *Systems* 6 (2): 18.

Holland, O. Thomas. 2018. "Foundations for the Modeling and Simulatino of Emergent Behavior Systems." In *Engineering Emergence: A Modeling and Simulation Approach*, p. 219. Boca Raton, FL: CRC Press.

Jackson, Daniel, and Craig A. Damon. 1996. "Elements of Style: Analyzing a Software Design Feature with a Counterexample Detector." *IEEE Transactions on Software Engineering* 22 (7): 484–495.

Maier, Mark W. 1998. "Architecting Principles for Systems-of-Systems." *Systems Engineering: The Journal of the International Council on Systems Engineering* 1 (4): 267–284.

Mason, Moya K. 2014. "Future Scenarios: The art of storytelling." https://www.moyak.com/papers/scenarios-future-planning.html.

NPS. n.d. *Monterey Phoenix Home*. Naval Postgraduate School. https://wiki.nps.edu/display/MP/Monterey+Phoenix+Home.

Oetsch, Johannes, Michael Prischink, Jorg Puhrer, Martin Schwengerer, and Hans Tompits. 2012. "On the Small-Scope Hypothesis for Testing Answer-Set Programs." In *Thirteenth International Conference on the Principles of Knowledge Representation and Reasoning*.

Ogilvy, Jay. 2015. "Scenario Planning and Strategic Forecaseting." *Stratfor*. https://www.forbes.com/sites/stratfor/2015/01/08/scenario-planning-and-strategic-forecasting/#ca9aa48411a3.

Orders in Council, Expedition in Virginia. 1974

Pagels, Heinz R. 1989. *Dreams of Reason*. New York: Bantam Books.

Thomas, Timothy. 2004. "Russia's Reflexive Control Theory and the Military." *Journal of Slavic Military Studies* 17: 237–256.

Wiggins, Warren. 2014. *War Game Adjudication: Adjudication Styles*. Newport, RI: United States Naval War College.

7 DEVS-Based Modeling and Simulation to Reveal Emergent Behaviors of IoT Devices

Moath Jarrah and Omar Al-Jarrah
Jordan University of Science and Technology

CONTENTS

7.1 INTRODUCTION

The concept of Internet of Things (IoT) refers to a system that connects billions of devices (things) to the Internet. These devices are called IoT devices, and they include different types of sensors, routers, printers, cameras, smart meters, smartwatches, home appliances, light bulbs, traffic lights, and many more. Each individual device is considered as a system on its own that aims at providing a service or functionality. The objective of connecting all devices to the Internet is to achieve optimization in terms of energy and resources management. Artificial intelligence and machine learning applications are deployed to process data and make decisions based on the features and characteristics of the IoT system. Individual devices and resources are managed automatically which leads to the vision of smart cities. Hence, the IoT system is a complex System of Systems (SoS) (for more discussion see Ref. [1]).

DOI: 10.1201/9781003160816-9

IoT devices are manufactured by many vendors. Some vendors, especially small businesses, produce devices with profit as the main motive without considering standardization in their designs. Connecting devices from different vendors to the Internet results in a heterogeneous distributed environment. A device is designed to serve specific functionalities. Additionally, companies focus on devices' simplicity and ease of use to attract customers. Hence, manufacturers rush to win the market without considering different risk factors and emergent behaviors that could result from having a large number of devices connected to the Internet backbone. Many users and system administrators rely on the simplicity of installation and do not configure the devices properly. For example, leaving remote login connections with the default username and password is not one of the best practices. Moreover, many devices that were connected to the Internet in the past are forgotten, and some devices will be forgotten in the future, leaving them without new updates and secured configurations. All of the aforementioned aspects increase the surface of threat and pose challenges to the functionality of IoT SoS.

Since 2016, a new type of attacks (IoT botnets) was discovered where hundreds of thousands of misconfigured IoT devices were infected. The emergent behaviors of these infected devices produced an unprecedented traffic volume of Distributed Denial of Service (DDoS) attacks. Service providers were not ready for such an aggregate Internet traffic that occurred because of changes in the behavior of thousands of IoT devices. This is likely to become worse as more and more devices are being connected to the Internet. A botmaster is the designer of the botnet, and the infected devices are called bots. The botmaster exploits the simplicity of devices and installs programs on bots that enable her/him to remotely control the bots. A command and control (C&C) server is used by the botmaster to send commands to the bots through a secured channel (for more details see Ref. [2]). The botnet design, devices simplicity, complexity, and large scale of the IoT SoS result in emergent behaviors of the IoT devices. Hence, there is a need for IoT designers and administrators to use modeling and simulation to compute the functionality and behaviors of IoT devices to prevent different emergent behaviors that could negatively (negative emergent behaviors) impact the IoT SoS performance.

Top-down or equation-based methods are too difficult to be used to evaluate and analyze complex systems such as the IoT SoS. Hence, a bottom-up modular approach is more suitable where a component-based modeling can be applied for subsystems and devices. A device is considered as a component model. Some devices share the same component model. Each model describes the expected (or normal) behavior of an IoT device. Multiple models can then be connected to a central processing model to virtually simulate a higher level (e.g., a smart home). Smart homes models can then be connected in a hierarchical approach to simulate a residential area, city, province, country, and so on. The behavioral characteristics of the smallest entities (IoT devices) are described using microscopic features. The overall IoT SoS characteristics are described using macroscopic features. For example, the traffic volume of the IoT SoS is a macroscopic feature, and a temperature sensor that reports data every 1 hour is a microscopic feature of the sensor. At the lower level, an IoT device is described using microscopic features that determine its states and behaviors as a result of interacting with its surrounding environment. At the higher level, the IoT system has macroscopic features that characterize its functionality and the service level agreement (SLA). Discrete Event System Specifications (DEVS) formalism can be used to simulate normal and abnormal traffic patterns to reveal emergent behaviors.

DEVS formalism is a powerful modeling and simulation tool that provides a systematic bottom-up approach. An atomic model in the DEVS environment represents the lowest component model. Multiple atomic models can be connected at a higher level which is called a coupled model. Multiple coupled models are connected together in a higher level which is called a hierarchical model. IoT devices can be modeled as atomic models. DEVS environment provides a timed sequence of events that cause a state of a model (e.g., device) to change during a given simulation.

This chapter explains how a modeling and simulation approach can be used to discover emergent behaviors that result from altering normal traffic patterns of different IoT devices in terms of a system's functionality. Network administrators can monitor traffic patterns since they have knowledge about the scope of their connected devices. Hence, administrators can segment or isolate subnetworks that have suspicious traffic activities. This chapter presents detection and controlling solutions that can be used to mitigate undesirable situations.

7.2 IoT DEVICES

The IoT era covers a wide range of connected devices such as household appliances, surveillance cameras, entertainment devices, lights, network routers, and different types of sensors. Billions of devices are connected to the Internet backbone in order to report data and information to servers. Hence, the behavior of a device is described based on its network traffic activities. Modeling different devices and simulating their traffic behaviors are essential in understanding the nature of the traffic found in IoT systems.

IoT devices are limited in their computational power and are designed for specific functionalities. For example, a temperature sensor is designed to periodically report the temperature of an environment to a central station or server. Some devices send data periodically or sporadically according to the interaction with the environment such as motion sensors and surveillance cameras. Generally, IoT devices are incapable of performing intensive computational tasks.

7.2.1 MICROSCOPIC AND MACROSCOPIC FEATURES OF AN IoT SYSTEM

In the context of this chapter, microscopic features of IoT devices refer to the traffic characteristics of a single IoT device such as type, volume, and periodicity. The macroscopic features are the traffic characteristics of the entire IoT system such as the aggregate traffic volume and rate. The modeler describes the behaviors of individual devices at the microscopic level. This might seem as a tedious task, given the number and diversity of devices. However, many types of devices share common behaviors and characteristics. Additionally, modeling a subset of IoT devices and simulating their behaviors provide insights and conclusions that can reveal important information about the IoT environment. IoT emergent behaviors can be revealed at the macroscopic level of the system.

Many studies have been considering applying time series techniques to model and detect network traffic for security purposes such as the work in Refs. [3–7]. Some of these studies have used the time series of the traffic pattern to infer important features from the IoT network traffic. A mathematical model of machine traffic (sensor networks) was studied by Anders Orrevad in Ref. [8]. The model followed similar concepts and characteristics as mentioned earlier in terms of low rate, bursts of traffic, and ON/OFF states. There are two methods that were addressed in literature on traffic modeling. The first one considered each device independently and modeled its traffic pattern. The second one modeled the aggregate traffic that is generated by many devices. The aggregate traffic model is imprecise as it simplifies the complexity of having a large number of devices [9]. On the other hand, modeling each device's traffic separately is easier and more precise. The latter is more attractive as it is a bottom-up approach. The aggregate traffic modeling occurs at the overall system level and results in approximating many details.

There are several studies that addressed the traffic characteristics of IoT devices. Devices send traffic in bursts and can be modeled by having ON and OFF states, where the traffic is sent in an ON state. The time spent in a state can be different based on the event that triggers the state transition. Figure 7.1 shows examples of traffic pattern bursts. The traffic types can be divided into two categories (signaling and event-driven). Many studies have focused on understanding the nature of the IoT traffic for security purposes. Table 7.1 shows different macroscopic characteristics of the IoT traffic. The traffic volume that is produced by a single IoT device is very small. However, the aggregate traffic volume of thousands of IoT devices is massive.

The authors of Refs. [10–12] have conducted experiments to understand the normal traffic behavior of different types of IoT devices. The authors found different attributes that were used to identify IoT devices. The attributes include active time, sleep time, number of protocols, number of ports, DNS requests and intervals, packets size, active traffic volume, flow rate, number of servers, and NTP interval. These attributes are microscopic features that describe IoT devices. They have also found that the sleeping time duration is less than 20 seconds most of the time (85%). The majority of IoT devices generate small bursts of traffic with a volume of less than 1KB. Additionally, the packet size, most of the time, is less than 500 bytes. HTTPS with TCP port 443 and HTTP with port 80 are the most dominant protocols that are used by IoT devices. UDP with ports 53, 123, and 1900 are among the protocols used by IoT devices, where UDP port 1900 is used in local networks. The sleep times for some devices were found to be 20, 60, and 90 seconds. The size of packets during an active time is found to be 114, 342, and 3341 bytes with an average packet size of 75, 200, and 225. Moreover, the authors have found that DNS requests of some IoT devices are sent to a few numbers of domains that occur every 5, 10, and 30 minutes. Most IoT devices send DNS requests to a small number of domain names (i.e., 10 domains). The reason for this behavior is that a device is pre-programmed to connect to the servers of the manufacturers or specific service providers. The network time protocol (NTP), which is a UDP protocol with the port number 123, is used frequently by IoT devices for time synchronization. The NTP requests were found by the authors to occur periodically. These findings adhere to studies that were previously published on understanding machine-type communication traffic such as

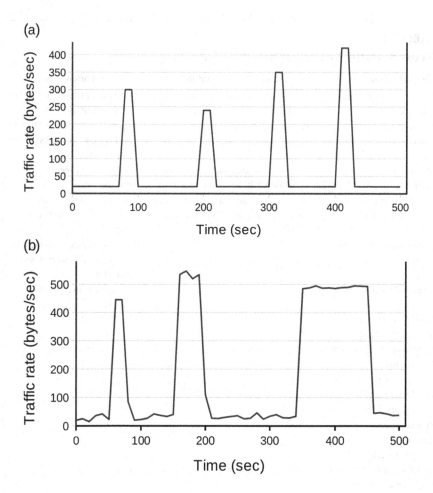

FIGURE 7.1 Bursts in IoT network traffic.

[9,13]. Hence, when modeling an IoT device, it is important to propose models that generate low traffic rates with protocols and port numbers that are similar to the aforementioned characteristics.

7.2.2 IoT Botnets and the Emergent Behavior

Botnets are networks of infected devices that are used by attackers to perform network malicious activities [2]. Previous studies have shown that the traffic pattern that is generated by botnets can be detected using flow and packet attributes [14–18]. IoT botnets, such as Mirai, Torii, Hajime, BASHLITE, QBot, and their variants, target IoT devices as they are weakly configured and can be infiltrated easily using manufacturer's default usernames and passwords. Also, IoT devices are always connected to the Internet, which make them attractive and reliable surface for hackers to launch network attacks. IoT botnets can infect millions of devices that are then used

TABLE 7.1

Network Traffic Characteristics of IoT Devices

General Behaviors	Traffic Volume and Rate	Networking Services
Periodic traffic to indicate alive device	Small traffic volume, mostly less than 500 K bytes per device per day.	Standard protocols such as TCP, UDP, ICMP, and IGMP.
Traffic can be unicast or multicast	Packet size is less than 500 bytes.	Standard ports such as DNS 53, NTP 123, ICMP 0, SSDP 1900, and TLS/SSL 443.
Destination can be local or external	Most devices peak load (when active) is about 1 KB. Few have 10 KB.	NTP requests every 60, 300, or 600 seconds.
Servers, protocols, Ports are deterministic	Maximum ON period (active time) for devices is less than 250 seconds.	Non-standard ports such as 33434, 56700, 8883, and 25050.
Few domain names and fixed		Devices of same manufacturer use same ports such as 8443 and 3478.
Users' interaction traffic		DNS queries every 5 or 30 minutes.
ON/OFF traffic pattern		
Most devices have deterministic sleep time		
Periodic time synchronization (NTP)		
Few server ports		
Non-standard ports		
Prominent domain names are used		
Many IoT devices keep their secured connection alive for a long period		

to launch a massive DDoS attacks such as the unprecedented attack on Dyn, OVH, and KrebsonSecurity which occurred in 2016 by Mirai IoT botnet.

Many studies have been focusing on understanding IoT botnets and the emergent behaviors of infected devices. The authors of Ref. [19] have found out that the Mirai botnet was used by many command and control servers to launch DDoS attacks on service providers. Table 7.2 shows the different attack types that were found. As noted by the authors, attackers (botmasters) send commands to all devices within the botnet. A command consists of the attack type, the target, the duration, and some options. See Ref. [20] for more details on the attacks, commands' options, and customizable fields. Additionally, the authors of the study in Ref. [21] have discussed that Mirai and other botnets are constantly targeting IoT devices because of their vulnerability and misconfiguration. IoT botnets are becoming more sophisticated in their design and attack strategies. Hence, countermeasures are urgently needed to detect them and isolate subnetworks that contain infected devices.

TABLE 7.2
Mirai DDoS Attacks

Attack Type	Example, Assuming a Domain Name Server X
HTTP flood	IoT devices send many HTTP requests to retrieve objects from domain X (IPsrc=bot, IPdest= server X, Port=80)
DNS flood	DNS reply messages from a massive number of DNS servers. Spoofed and non-spoofed IP address
SYN flood	IoT devices send many SYN packets that are destined to server X
ACK flood	IoT devices send many ACK packets that are destined to Server X
UDP flood (or plain)	IoT devices send many UDP packets that are destined to Server X
VSE flood	IoT devices flood server X with a large number of VSE queries
GRE-IP flood	IoT devices send payload/data to server X. Data is encapsulated within large GRE packets
ACK-STOMP flood	IoT devices flood server X with fake STOMP requests leading to network saturation and server's resource exhaustion
GRE-ETH flood	Like GRE-IP flood but with Ethernet encapsulation

The traffic behavior of infected IoT devices (or bots) is different than their normal behavior that is discussed in Section 7.1.2. Hundreds of thousands and even millions of infected IoT devices can be used to launch short-term or long-term DDoS attacks. Short-term attacks can be in the range of few minutes, and long-term attacks are in an order of hours.

7.2.3 DISCRETE EVENT SYSTEM SPECIFICATION (DEVS) FORMALISM

DEVS formalism is a powerful tool that supports modeling specifications and a simulation environment [22]. DEVS has been used in many applications such as the modeling and simulation of smart grid renewable components [23] and smart home devices [24]. A system's individual components at the microscopic level are described as atomic models in DEVS, as shown on the top left of Figure 7.2. Two or more atomic models are connected together in a higher level which is called a coupled model (top right of Figure 7.2). Atomic and coupled models are connected together in a hierarchical model which is the highest level (macroscopic level) in DEVS, which is shown at the bottom of Figure 7.2.

7.3 MODELING IoT DEVICES

The traffic behavior of physical IoT devices at the microscopic level (DEVS atomic models) is modeled based on the normal behavior. This is required for the simulation in order to reveal emergent behaviors [25]. Hence, uniform distribution is used to model periodic traffic patterns and Poisson distribution is used to mimic the environment interactions with IoT devices. The intervals that are used for the uniform distribution are close to the values that are reported for IoT devices. Moreover, the interactions occur less frequently than periodic traffic, and hence, a small arrival

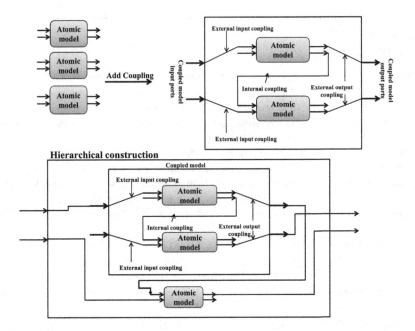

FIGURE 7.2 DEVS models.

rate parameter (λ) of the Poisson arrivals is used. The widely known algorithm by Donald Knuth [26,27] for Poisson distribution is used as shown in Algorithm 7.1. The number of user and/or environment interactions with devices that are used is approximately 340 interactions per day. This number can be changed by the modeler to increase or decrease the interactions per day by using different values for the parameter λ in Algorithm 7.1.

Algorithm 7.1 Donald Knuth Poisson Generator

1. $L \leftarrow \exp(\lambda); K \leftarrow 0; P \leftarrow 1$
2. $Do \ \{K++$
3. $u \leftarrow U[0,1]$
4. $P \leftarrow P \times u$
5. $\}while \ (P > L)$
6. $return \ K - 1$

7.3.1 Modeling Normal Behaviors

This section considers the microscopic features of an IoT device at the lowest level which consist of its states and traffic behavior. A DEVS atomic model of an IoT device transits between two states (ON and OFF). During the ON state, a message/packet is

TABLE 7.3
Typical Packet Size

Packet Type	Typical Size (bytes)
HTTPS/HTTP	75, 114, 200, 225, 342, 3341[a]
DNS	60
SYN	60 or 72
ACK	60 or 72
SSDP protocol	Between 110 and 330
NTP	Between 48 and 64

[a]Based on the findings of the authors in Refs. [10–12].

sent. The packets can be DNS, SYN, HTTPS, HTTP, ACK, UDP port 1900 (SSDP protocol), or NTP. Table 7.3 shows the typical sizes of each packet. DEVS modeling and simulation formalism allow us to generate different simulations for different packet sizes and time periods of the ON state. For examples, the authors of Refs. [10–12] have found out that IoT devices send DNS queries every 300, 600, and 1800 seconds. Also, NTP requests are sent every 50, 300, and 600 seconds. The presented modeling approach of this section is to have a mix between these time steps to mimic physical IoT devices normal behaviors. Other observations such as the traffic fraction of HTTPS, HTTP, and SSDP protocol packets were reported in research. These fractions can be scaled to produce similar results. For example, if 10% of the traffic is NTP (during a time period of a normal behavior), then the modeler generates an NTP traffic that contributes to 10% of the total traffic by having the NTP-ON period to be 10% of the ON periods of the total network traffic. User interactions with physical IoT devices result in an immediate transition to the ON state to report data (typically by sending DNS requests to connect to a server).

Figure 7.3 shows a sketch diagram of seven IoT atomic models within a coupled model (*Collection Model*) in DEVS modeling and simulation environment. A user or environment interaction with the devices is triggered by the model (*Interaction Model*) that causes an immediate state transition for *IoT_D3* model. The *Analyzer* model logs traffic and performs analysis as we will see in Section 7.4.

7.3.2 Modeling IoT Botnet Behaviors

As noted by the studies in Refs. [19–21], Mirai IoT botnet has launched short and long-lived DDoS attacks. The short-lived attacks were approximately for 25 seconds, and the long-lived attacks were for 1, 5, and 10 hours. Also, it was found that SYN and ACK floods were among the top DDoS attacks that were used. The same basic principles apply to other IoT botnets. When a hacker (botmaster) sends a command to an IoT device, the command contains the type of network packets to be sent and the time duration. In modeling the IoT botnet attack behavior, a modeler can assume different scenarios and run simulations in order to reveal insights and understand

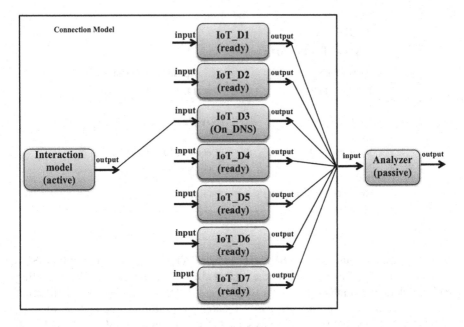

FIGURE 7.3   IoT devices models in DEVS formalism.

the traffic behaviors. Equation 7.1 shows the maximum number of packets that an IoT device i can send during 1 second, where $Link_{Bi}$ is the allocated link bandwidth for device i in bits per second (bps), 8 is the size of a byte in bits, and P_S is the packet size in bytes. For example, if the Internet link speed is 100 Mbps, 10% is allocated for device i, and the packet size is 64 bytes, then the maximum number of packets that device i can send is 19,531 packets per second. Based on Section 7.2.1, the average packet size for IoT devices can be assumed to be 200 bytes. Hence, during a 1 second, on average, the maximum packets that can be sent is 6250 packets, given that the allocated bandwidth is 10% of a 100 Mbps link. If the actual number of packets that an IoT device sends is 60% of the maximum number, then a device can send 3750 packets per second. The following simulations consider that the number of packets that are sent by an IoT device during a DDoS attack period is 60% of the maximum number. Simulations enable us to explore different parameters and settings in order to understand different scenarios.

$$N_{max} = \frac{Link_{Bi}}{8 \times P_S} \tag{7.1}$$

Figure 7.4 shows a sketch of a DDoS attack atomic model that is added to the coupled model (*Collection Model*). The *DDoS Attacks* model provides the functionality of a botmaster (or a command-and-control server) which sends a command to IoT devices specifying the type of attack (i.e., DNS, SYN, etc.) and the duration. The top-left box in the figure shows four durations which are short-lived duration (25 seconds) and long-lived (1, 5, and 10 hours) attacks.

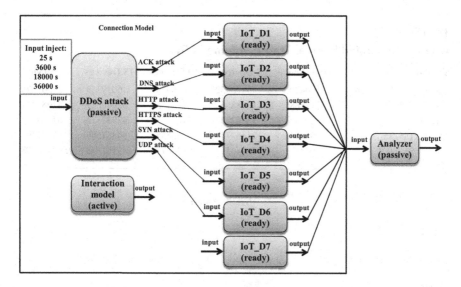

FIGURE 7.4   DDoS attack model within the IoT devices coupled model.

7.4 TRAFFIC ANALYSIS

The following sections use DEVS modeling and simulation to study and analyze the behavior of multiple IoT devices for different time periods. The time unit corresponds to 1 second which is based on the modeling approach that was done in Section 7.3.

7.4.1 Normal Behavior Identification

Table 7.4 shows the traffic volume (macroscopic feature) that is generated by different numbers of IoT devices based on their normal behaviors. The volume is the average of five runs of the simulation. An IoT device can generate a volume of about 30 MB during a week. This volume can be monitored by a household device or network administrators to keep track of the devices within their subnetworks. Figure 7.5 shows a typical sample of a device's traffic pattern. The figure shows the nature of the traffic bursts of the normal behavior which adheres to the findings of many studies as mentioned earlier. The average packet size of 18 hours is about 1.1 KB. However, we can see from the figure that there are packets of more than 14 KB which indicate payload data.

Another important macroscopic feature of IoT systems is the traffic flow rate. The flow rate can give network administrators an indication of how the bandwidth is being exploited by devices. It is computed by dividing the number of bits that are sent by the time period (in seconds). Some devices such as high-resolution surveillance cameras reserve larger bandwidths than other devices (e.g., temperature sensors) in order to perform their functionalities appropriately. In such cases, a high-speed network link is required, and more packets can be sent in 1 second according to Equation 7.1 as we will see in Section 7.4.2.

TABLE 7.4

Average Generated Normal Traffic Volume in Bytes

Simulation time	One Device	5 Devices	10 Devices	15 Devices	20 Devices
6 hours (21,600 seconds)	1064200	5445032	10669799	16039359	21301676
12 hours (43,200)	2178886	10788862	21451859	32040227	42983371
18 hours (64,800)	3267457	15862933	31861751	47937815	64449387
24 hours (86,400)	4369348	21487774	42525553	64351569	85785198
1 week (604,800)	29942471	149446521	299436779	449101348	599498408

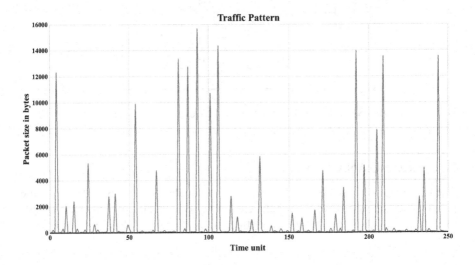

FIGURE 7.5 Normal traffic pattern of an IoT model.

Table 7.5 shows the flow rate of multiple devices. During a normal behavior, an IoT device has a flow rate of about 0.4 Kbps, and 20 devices have 7.9 Kbps. Table 7.6 shows the average flow rate per device which indicates that an IoT sends a low bit rate during a normal behavior.

7.4.2 ABNORMAL BEHAVIOR IDENTIFICATIONS

IoT devices are vulnerable, and hackers exploit their simplicity to infect them with malware (bot programs). The large number of IoT devices allows for launching massive DDoS attacks. In this section, we consider scenarios where some IoT devices can send a higher number of bytes per second in case of UDP flood and HTTP flood (e.g., they get higher bandwidth or have higher bps network speed such as surveillance cameras that have the capacity to send large packets which require a higher bandwidth). GRE IP flood attacks are assumed to behave similar to the HTTP flood attacks by carrying a large payload data. In addition, ACK and SYN flood attacks have similar packet sizes and macroscopic traffic features.

TABLE 7.5

Average Normal Aggregate Flow Rate in bps

Simulation time	One Device	5 Devices	10 Devices	15 Devices	20 Devices
6 hours	394.1481481	2016.678519	3951.777407	5940.503333	7889.50963
12 hours	403.4974074	1997.937407	3972.566481	5933.37537	7959.883519
18 hours	403.3897531	1958.38679	3933.549506	5918.248765	7956.714444
24 hours	404.5692593	1989.608704	3937.551204	5958.478611	7943.073889
1 week	396.0644312	1976.805833	3960.803955	5940.494021	7929.873122

TABLE 7.6

Normal Flow Rate Per Device in bps

6 hours	394.1481481	403.3357037	395.1777407	396.0335556	394.4754815
12 hours	403.4972593	399.5874815	397.2566481	395.558358	397.9941759
18 hours	403.3897531	391.677358	393.3549506	394.5499177	397.8357222
24 hours	404.5692222	397.9217407	393.7551204	397.2319074	397.1536944
1 week	396.0644233	395.3611667	396.0803955	396.0329347	396.4936561

TABLE 7.7

Traffic Volume in Bytes for a Short-Lived DDoS Attack during a 1-Hour Simulation

	SYN Flood	UDP Flood	HTTP or HTTPS or GRE IP Flood
Normal traffic	171456	171456	171456
1 attack	6365739	20762824	655612960
5 attacks	31093198	103038051	3267579566
10 attacks	62073813	205991876	6541773113
15 attacks	92981379	308815067	9820227770
20 attacks	123981222	411724530	13098277343

A botmaster sends commands to the infected IoT devices that contain the packet size. The work in Ref. [20] has shown that a typical packet size of 512 bytes can be used by DDoS attacks. However, the botmaster can send commands to the bots (infected IoT devices) that carry a different size of packets. Hence, the modeling and simulation environment enables us to randomize the size of the packets in the range of 48–3400 bytes to reveal their impact on the network. The packets' size is randomly selected based on Table 7.3.

Based on the studies in Refs. [19–21], different simulations for short-lived and long-lived types of attacks for one device can be conducted. Tables 7.7 and 7.8 show the macroscopic traffic volume (in bytes) of short-lived (25 seconds) DDoS attacks

TABLE 7.8

Traffic Volume in Bytes for a Combination of Short-Lived DDoS Attacks during a 1-Hour Simulation

	SYN, UDP, and HTTP Attacks	Fraction of Normal Traffic
Normal traffic	171456	1
3 attacks	680179144	3967.076941
6 attacks	1358361332	7922.506836
9 attacks	2040270944	11899.67656
12 attacks	2731656639	15932.11459
15 attacks	3396830891	19811.677

TABLE 7.9

Flow Rate in Kbps for Short-Lived DDoS during a 1-Hour Simulation

	SYN Flood	UDP Flood	HTTP or HTTPS or GRE IP Flood
Normal traffic	0.381013333	0.381013333	0.381013333
1 attack	14.14608667	46.13960889	1456.917689
5 attacks	69.09599556	228.9734467	7261.287924
10 attacks	137.9418067	457.7597244	14537.27358
15 attacks	206.6252867	686.2557044	21822.72838
20 attacks	275.5138267	914.9434	29107.28298

during a 1-hour time-period of the simulation. The first row is the average volume of the normal behavior (without DDoS attacks). When a botmaster sends commands to the infected IoT devices to launch DDoS, a large difference occurs in the traffic volume that is generated by the device. The background traffic (normal traffic) becomes negligible in comparison with the generated traffic as a result of the attack. During a normal behavior, IoT devices spend most of time in the OFF state where no traffic is generated.

Tables 7.9 and 7.10 show the flow rate of the short-lived DDoS attacks for one device launching the different types of attacks. As shown in Table 7.9, even one small SYN flood attack results in more than 30 times increase in the IoT traffic flow; and since botmasters launch multiple number of attacks, the fraction of increase in the flow rate is in an order of magnitude. Hence, system administrators need to be equipped with the tools to monitor IoT devices as they might produce traffic with unique characteristics that enable us to detect their abnormal behaviors.

The following results are for long-lived DDoS attacks. Tables 7.11 and 7.12 show the traffic volume and flow rate for a 1-hour DDoS attack during a 6-hour simulation. From Table 7.12, we can notice that the smallest attack (SYN flood) results in an increase in the flow rate of more than 830X the normal traffic bit rate. The 1-hour-long campaign of the combination of attacks results in more than 92,000X increase in the flow rate.

TABLE 7.10

Flow Rate in Kbps of a Combination for Short-Lived DDoS during a 1-Hour Simulation

	SYN, UDP, and HTTP Attacks	Fraction of Normal Traffic
Normal traffic	0.381013333	1
3 attacks	1511.509209	3967.076941
6 attacks	3018.580738	7922.506836
9 attacks	4533.935431	11899.67656
12 attacks	6070.348087	15932.11459
15 attacks	7548.513091	19811.677

TABLE 7.11

Traffic Volume in Bytes for a Long-Lived DDoS during a 6-Hour Simulation

	SYN Flood	UDP Flood	HTTP	A Combination of HTTPS, SYN and UDP
Normal traffic	1064200	1064200	1064200	1064200
One-time attack	892033624	2964357589	94224530044	98087464969

TABLE 7.12

Flow Rate in Kbps for a Long-Lived DDoS during a 6-Hour Simulation

	SYN Flood	UDP Flood	HTTP	A Combination of HTTPS, SYN, and UDP
Normal traffic	0.394148148	0.394148148	0.394148148	0.394148148
One-time attack	330.3828237	1097.910218	34897.97409	36328.69073

The last simulation considers the scenario where multiple devices launch attacks. Table 7.13 shows a simulation of 20 IoT devices system. Five devices launch short-lived, 1-, and 2-hour DDoS attacks. The attacks are combinations of SYN, ACK, HTTPS, HTTP, and UDP. This simulation shows the flow rate (bits per second) that are likely to be seen in the network. The attack results in about 1.1 Gbps. Given the fact that billions of IoT devices are going to be connected to the Internet backbone, and hackers will always find ways to exploit their security to form IoT botnets, then it is likely that the flow rates of IoT botnet DDoS attacks are going to reach to values in an order of Tbps. Such a tremendous bit rate can cripple the Internet services and cause a massive damage to organizations. Hence, IoT devices should not be connected directly to a home network or an organization unless they are continuously monitored and configured correctly. We recommend that manufacturers should produce special-purpose monitoring devices. The IoT devices get connected to the

TABLE 7.13
Traffic Volume in MB for Five Devices in a 20 Devices System

Attack	Volume (MB)
Short-lived	3438.542766
1-hour	494893.3896
2-hour	989732.8602

Internet through the special-purpose device which continuously monitors the flow rate. If the flow rate becomes higher than a threshold (based on the normal behavior), the special-purpose device disconnects the IoT devices and raises an alarm for administrators, so they inspect the devices.

7.5   IoT MODELING USING MONTEREY PHOENIX

Monterey Phoenix* is a modeling tool that provides different system behaviors and scenarios based on abstract modeling rules. It has its own language syntax to describe a system of interest. System components are modeled, and the different interaction behaviors between the components are generated. Hence, it is a useful tool to reveal emergent behaviors. Figure 7.6 illustrates different interaction of IoT devices with a network and an environment. An interaction with the environment has many examples such as receiving a command from a user, sensing a temperature or humidity, sensing a movement, and so on. When the interaction with the environment occurs, an IoT device responds by sending network traffic that carries information or data as explained in Section 7.3. Figure 7.6a illustrates such a behavior. Additionally, an IoT device is scheduled to send periodic traffic to synchronize its time clock and to inform other network devices about its existence which is shown in the state "Sync" in the figure.

Figure 7.6b shows two interaction scenarios with the environment which are marked in red color (1 and 2). The first interaction (1) causes the IoT device to transit into the "Active" state to send traffic to the destination through the network channel which conforms with its expected functionality. However, in the second scenario (2), the IoT transits into "Sync" state which could happen in reality as IoT devices are scheduled to perform periodic tasks (i.e., synchronize the time clock). If the IoT device reports the data as a result of the interaction with the environment after it finishes its scheduled task, then the expected functionality of the device is completed. This is called a "positive emergent behavior" as it does not affect its normal behavior and the goal is achieved. However, if the IoT device does not report the data for some reason such as running out of battery, a hardware failure, or any other types of failures, then its expected functionality is incomplete, and this results in a "negative emergent behavior." To solve this situation, redundancy (multiple devices that perform the same task) is important when sensing a critical environment. On another hand, Figure 7.6c shows how the DDoS attack occurs. In this scenario, the botmaster

---

* https://firebird.nps.edu

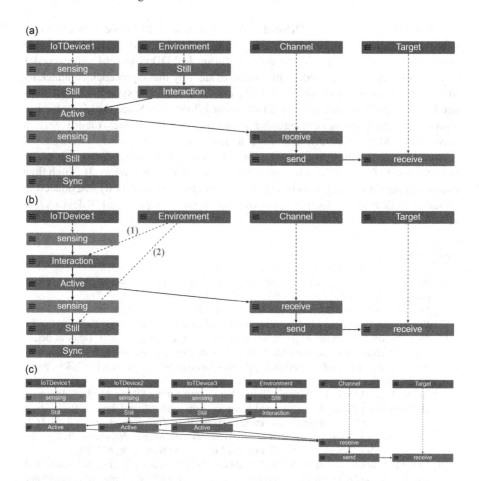

FIGURE 7.6 IoT devices model in Monterey Phoenix. (a) Normal behavior (b) Emergent behavior (c) DDoS attack behavior

interacts with many IoT devices by sending a command to lunch one of the DDoS attacks. The IoT devices respond by transiting into "Active" state to send traffic to the target. When many devices are involved in this behavior, the target will be overwhelmed with the amount of incoming traffic which results in crippling its services.

7.6 CONCLUSION

This chapter described a bottom-up modeling and simulation method based on microscopic and macroscopic of IoT systems. The normal traffic behavior of IoT devices is modeled at the lower level as atomic models in the DEVS formalism. Also, the DDoS attacks of IoT botnets were modeled in DEVS environment. DEVS modeling and simulation environment allowed us to perform different scenarios of normal and abnormal network traffic generation. Based on the simulation results, a small SYN flood attack can result in more than 30 times increase in an IoT traffic flow rate. Moreover, a five-device DDoS attack in a system of 20 IoT devices can result a flow rate of more than 1 Gbps. Given that billions of IoT devices will be available and connected to the Internet, the order of future IoT DDoS attacks will exceed 1 Tbps by an order of magnitude. This is also evident by the unprecedented attack by Mirai IoT botnet which exceeded 600 Gbps in 2016. Hence, a system administrator needs to be equipped with right tools to monitor IoT devices. Since the IoT traffic has unique characteristics that distinguish it from other non-IoT devices, the IoT devices network should be isolated from non-IoT devices. We suggest that IoT devices should be connected to the Internet through a special-purpose monitoring device that maintains the level of the flow rate to match the devices' normal behaviors. If a high flow rate is detected, the special-purpose device reacts instantaneously by disconnecting its subnetwork from the Internet and raises an alarm to inspect the IoT devices that belong to the subnetwork.

REFERENCES

[1] Johan Lukkien. A systems of systems perspective on the internet of things: Invited paper. *ACM SIGBED Review*, 13(3):56–62, 2016.

[2] Georgios Kambourakis, Marios Anagnostopoulos, Weizhi Meng, and Peng Zhou. *Botnets: Architectures, Countermeasures, and Challenges.* CRC Press: New York, 2019.

[3] Anaël Bonneton, Daniel Migault, Stephane Senecal, and Nizar Kheir. Dga bot detection with time series decision trees. In 2015 4th *International Workshop on Building Analysis Datasets and Gathering Experience Returns for Security (BADGERS)*, pages 42–53, 2015.

[4] Abbas Acar, Hossein Fereidooni, Tigist Abera, Amit Kumar Sikder, Markus Miettinen, Hidayet Aksu, Mauro Conti, Ahmad-Reza Sadeghi, and Selcuk Uluagac. Peek-a-boo: I see your smart home activities, even encrypted! In *Proceedings of the 13th ACM Conference on Security and Privacy in Wireless and Mobile Networks, WiSec '20*, pages 207–218. Association for Computing Machinery: New York, NY, 2020.

[5] Seyyed Meysam Tabatabaie Nezhad, Mahboubeh Nazari, and Ebrahim A. Gharavol. A novel DoS and DDoS attacks detection algorithm using ARIMA time series model and chaotic system in computer networks. *IEEE Communications Letters*, 20(4):700–703, 2016.

[6] Mbulelo Brenwen Ntlangu and Alireza Baghai-Wadji. Modelling network traffic using time series analysis: A review. In *Proceedings of the International Conference on Big Data and Internet of Thing, BDIOT 2017*, pages 209–215. Association for Computing Machinery: New York, NY, 2017.

[7] Mikel Izal, Daniel Morató, Eduardo Magaña, and Santiago García-Jiménez. Computation of traffic time series for large populations of IoT devices. *Sensors*, 19(1):78, 2018.

[8] Anders Orrevad. M2m traffic characteristics: When machines participate in communication. In *Information and Communication Technology*, KTH Sweden: Stockholm, 2009.

[9] Markus Laner, Philipp Svoboda, Navid Nikaein, and Markus Rupp. Traffic models for machine type communications. In *ISWCS 2013; The Tenth International Symposium on Wireless Communication Systems*, pages 1–5, 2013.

[10] Arunan Sivanathan, Daniel Sherratt, Hassan Habibi Gharakheili, Adam Radford, Chamith Wijenayake, Arun Vishwanath, and Vijay Sivaraman. Characterizing and classifying IoT traffic in smart cities and campuses. In *2017 IEEE Conference on Computer Communications Workshops (INFOCOM WKSHPS)*, pages 559–564, 2017.

[11] Arunan Sivanathan, Hassan Habibi Gharakheili, Franco Loi, Adam Radford, Chamith Wijenayake, Arun Vishwanath, and Vijay Sivaraman. Classifying IoT devices in smart environments using network traffic characteristics. *IEEE Transactions on Mobile Computing*, 18(8):1745–1759, 2019.

[12] Arunan Sivanathan, Hassan Habibi Gharakheili, and Vijay Sivaraman. Inferring IoT device types from network behavior using unsupervised clustering. In *2019 IEEE 44th Conference on Local Computer Networks (LCN)*, pages 230–233, 2019.

[13] Navid Nikaein, Markus Laner, Kaijie Zhou, Philipp Svoboda, Dejan Drajic, Milica Popovic, and Srdjan Krco. Simple traffic modeling framework for machine type communication. In *ISWCS 2013; The Tenth International Symposium on Wireless Communication Systems*, pages 1–5, 2013.

[14] Florian Tegeler, Xiaoming Fu, Giovanni Vigna, and Christopher Kruegel. Botfinder: Finding bots in network traffic without deep packet inspection. In *Proceedings of the 8th International Conference on Emerging Networking Experiments and Technologies, CoNEXT '12*, page 349–360, New York, NY, USA, 2012. Association for Computing Machinery.

[15] David McGrew and Blake Anderson. Enhanced telemetry for encrypted threat analytics. In *2016 IEEE 24th International Conference on Network Protocols (ICNP)*, pages 1–6, 2016.

[16] Blake Anderson and David McGrew. Identifying encrypted malware traffic with contextual flow data. In *Proceedings of the 2016 ACM Workshop on Artificial Intelligence and Security, AISec '16*, page 35–46, New York, NY, USA, 2016. Association for Computing Machinery.

[17] Bushra A. AlAhmadi, Enrico Mariconti, Riccardo Spolaor, Gianluca Stringhini, and Ivan Martinovic. Botection: Bot detection by building markov chain models of bots network behavior. In *Proceedings of the 15th ACM Asia Conference on Computer and Communications Security, ASIA CCS '20*, page 652–664, New York, NY, USA, 2020. Association for Computing Machinery.

[18] Basil AsSadhan and José M.F. Moura. An efficient method to detect periodic behavior in botnet traffic by analyzing control plane traffic. *Journal of Advanced Research*, 5(4):435–448, 2014. Cyber Security.

[19] Manos Antonakakis, Tim April, Michael Bailey, Matthew Bernhard, Elie Bursztein, Jaime Cochran, Zakir Durumeric, J. Alex Halderman, Luca Invernizzi, Michalis Kallitsis, Deepak Kumar, Chaz Lever, Zane Ma, Joshua Mason, Damian Menscher, Chad Seaman, Nick Sullivan, Kurt Thomas, and Yi Zhou. Understanding the mirai botnet. In *Proceedings of the 26th USENIX Conference on Security Symposium, SEC'17*, page 1093–1110, USA, 2017. USENIX Association.

[20] Chad Seaman. Threat Advisory: Mirai Botnet. https://www.akamai.com/uk/en/multimedia/documents/state-of-the-internet/akamai-mirai-botnet-threat-advisory.pdf, 2016. Online; accessed 23 December 2020.

[21] Constantinos Kolias, Georgios Kambourakis, Angelos Stavrou, and Jeffrey Voa. DDoS in the IoT: Mirai and other botnets. *Computer*, 50(7):80–84, 2017.

[22] Bernard P. Zeigler, Alexandre Muzy, and Ernesto Kofman. *Theory of Modeling and Simulation: Discrete Event; Iterative System Computational Foundations*. Academic Press, Inc., Cambridge, MA, 3rd edition, 2018.

[23] Moath Jarrah. Modeling and simulation of renewable energy sources in smart grid using DEVS formalism. *Procedia Computer Science*, 83:642–647, 2016.

[24] Majeda Albataineh and Moath Jarrah. DEVS-IoT: Performance evaluation of smart home devices network. *Multimedia Tools and Applications*, 80(11), 16857–16885, 2020.

[25] Stig Bosmans, Siegfried Mercelis, Peter Hellinckx, and Joachim Denil. Towards evaluating emergent behavior of the internet of things using large scale simulation techniques (wip). In *Proceedings of the 4th ACM International Conference of Computing*

for Engineering and Sciences, ICCES'18, New York, NY, USA, 2018. Association for Computing Machinery.

[26] Donald E. Knuth. *The Art of Computer Programming, Volume 4A: Combinatorial Algorithms, Part 1*. Addison-Wesley Professional, Pearson, Boston, MA, 2011.

[27] Donald E. Knuth. *Art of Computer Programming, Volume 2: Seminumerical Algorithms*. Addison-Wesley Professional, Pearson, Boston, MA, 1997.

8 Analyzing Emergence in Biological Neural Networks Using Graph Signal Processing

Kevin Schultz, Marisel Villafañe-Delgado,
Elizabeth P. Reilly, and Anshu Saksena
Johns Hopkins University Applied
Physics Laboratory (JHUAPL)

Grace M. Hwang
Johns Hopkins University Applied
Physics Laboratory (JHUAPL)
Kavli Neuroscience Discovery Institute
Johns Hopkins University

CONTENTS

DOI: 10.1201/9781003160816-10

8.1 INTRODUCTION

Emergent behavior is found throughout nature, including in biological neural networks where individual neurons connect and interact to form complex thought processes and task performance. We define *emergence* of a system to be the existence of complex system-level behavior that is not present in the much simpler individual entities of the system. The behavior of individual entities is locally defined and not directly tied to the system-level behavior. For instance, a bird in a flock might fly with the objective of being within a certain distance of approximately the six nearest birds, and yet, when viewed as a whole, this flock of birds can form ever-changing patterns with no centralized control.

In this chapter, we consider two different, yet related forms of emergence in biological neural networks. The first is the coordinated activity of neurons such as activity "bumps" with sustained and/or localized activity or synchronizing oscillations. These can be both self-sustaining, in the sense that the activity persists in the absence of stimuli, or transient behaviors brought about by a particular sequence of stimuli and vanish or decay after. These phenomena are often studied via relatively simple dynamical models of neurons interacting through a *static* network topology. These emergent behaviors are generally quite sensitive to both the parameters of the neuronal dynamics and the network interconnections between neurons. How, then, can large, complex neural networks form in a way that is highly dependent upon the experiences (i.e., stimuli) of the organism in question? The answer to this leads to the second form of emergence we consider. The process of network formation in a neural network over time is known as plasticity or learning. In this process, the connections between neurons (i.e., synapses) are strengthened or weakened over time based on the activity between the neurons in question. Thus, we have a dependency loop between two time scales of scales of emergence: The structure of the network is key to the emergent collective activity from individual neurons, and correlations in the neuronal activity drive the formation of the network.

In line with the overall theme of this volume, it is instructive to consider just how much the capabilities of an organism's brain far outstrip the capabilities of an individual neuron, or even a collection of neurons serving as a functional component (i.e., the sub-systems of the system of systems (SoS) we call the brain). As we will discuss below, an individual neuron is often modeled using a first-order differential equation and communicates with a single-bit channel (i.e., a spike train) to its neighbors in the network. Despite this apparent simplicity, modest collections of hundreds of thousands of neurons can form networks that serve essential navigational functions such as head direction and position estimation as part of the hippocampal formation in mammals, including humans (Poulter, Hartley, and Lever 2018). These are a few of the functions of the hippocampal formation, which itself interacts with many other brain regions and comprises a small fraction of the total number of neurons present in the brain. Thus, it is quite clear that the overall dynamics of the brain can be viewed as a complex SoS with emergent behaviors that are driven by the interactions of different functional components. These functional components, in turn, have emergent behaviors that are driven by the interactions of the individual neurons over their network interconnections.

Given that we have cast these emergent phenomena as a set of interacting processes on a network topology, we propose that graph signal processing (GSP) will serve as a useful analysis tool here. GSP builds on its roots in spectral graph theory (Chung and Graham 1997) and algebraic signal processing (Puschel and Moura 2008) to generalize techniques from classical signal processing to signals defined on irregular domains by graphs (Sandryhaila and Moura 2013; Shuman et al. 2013). Here, we propose that the outputs (or other state) of the neurons should be treated as the signal of interest with the network interconnections defining the graph structure. GSP techniques have been applied across various non-invasive neuroimaging techniques, including functional magnetic resonance imaging (fMRI) and electroencephalography. Goldsberry et al. (2017) and Medaglia et al. (2018) introduced GSP analysis for the joint study of structure and functional data from diffusion tensor imaging and fMRI, respectively, studying the alignment between functional activity and anatomical network in cognitive flexibility.

The study of emergence in biological neural networks is relevant to a broad range of application areas. Obviously, understanding the collective behavior of neurons is of fundamental importance in biology and medicine. Additionally, neural networks are the backbone of many advances in machine learning (Schmidhuber 2015) and are the inspiration for neuromorphic computing (Furber 2016), both of which hold further promise for continuing to revolutionize computation. Finally, to directly tie neural networks to a classic area of emergence, we note that neuronal dynamics have served as both a motivating analogy in swarm intelligence (Trianni et al. 2011) and also as a direct source of swarm dynamics (Monaco et al. 2020). To this last point, we note that graph theoretical techniques have been widely applied in the analysis of swarming dynamics (e.g., Tanner, Jadbabaie, and Pappas 2007), further reinforcing the use of GSP as an analytical tool for emergence.

In the following, we first introduce some basic models from computational neuroscience that will serve as the signals in our GSP analysis. Next, we review the basics of GSP and discuss in some detail the unique challenges presented by biological constraints. We then move to the discussion of short-term emergence in neural networks in the form of collective activity and analyze an example emergent phenomena in the form of spontaneous collective firing in a structured network. Next, we introduce the basics of network formation through plasticity and discuss some relevant examples of emergence that tie the collective dynamics of neurons to the formation of networks. We then consider a specific example of network formation that leads to a similar structured network previously considered and analyze its long-term behaviors using GSP. We conclude with discussion and lessons learned.

8.2 BIOLOGICAL NEURAL NETWORKS

The human brain is composed of about 85 billion neurons (nerve cells) and 85 billion glia cells. For decades, neurons were thought to be exclusively involved in the computation of the brain through their electrical properties. Emerging evidence reveal that chemicals, or more specifically neuromodulators, are involved in neuronal computation and that glia cells, in particular at least one form of glia cells, the microglial can produce negative feedback similarly to the role of inhibitory neurons (Badimon et al. 2020). In this chapter, a simplified mathematical abstraction of a biological neuronal network is described in the context of emergence. Thus details of neuromodulators, glia cells, and non-linearities within a neuron exhibited at particular dendritic branches (Gidon et al. 2020) that reveal how a single neuron can itself be modeled as a neuronal network capable of achieving many logical computations – AND, OR, XOR – will be ignored. Instead, we introduce some high-level and general concepts about various systems and processes in biological neuronal networks. To achieve this, we pursue the discussion at the level abstraction appropriate for computational or theoretical neuroscience; that is, at the level of relatively simple equations and interactions. For a more in-depth introduction into computational neuroscience, we refer the readers to full texts such as Dayan and Abbott (2001) or Miller (2018). For an accessible overview of a self-organized viewpoint of cognition, see Buzsáki (2019).

8.2.1 THE NEURON

The neuron is arguably the fundamental processing unit of biological neuronal networks and artificial neural networks (ANN). However, to speak of "the neuron" does a disservice to the breadth of form and function of neurons that occur in biology. At its core, a biological neuron is a cell with branches of dendrites (inputs), connected to a soma or cell body (processing), which is connected to one axon (output) that can transmit neural activity (information) to a few thousand other downstream neurons via synapses. Dendrites, which are typically covered by synapses, can extend far from the soma by hundreds of microns to receive inputs via synapses from many upstream neurons. Dendrites have been shown to critically contribute to the non-linear computations performed by neurons (Poirazi and Papoutsi 2020; Gidon et al. 2020). The soma is connected to one axon, that can extend over 1 m in humans eventually branching out, to potentially transmit neural activity to other neurons. Each biological neuron has a state-dependent, adaptive, electrical threshold that when exceeded allows neural activity to propagate through the axon to other downstream neurons. This transmission is done via an electro-chemical process in the terminus of the axon that releases tiny vesicles containing neurotransmitters that can lead to a change in the postsynaptic membrane voltage. One can think of the adaptive threshold of a real biological neuron to be an emergent property within the neuron. In this chapter, single-compartment, point-like neuron models will be discussed in the context of emergence; these models are devoid of any intracellular compartments (e.g., dendrites, axons, vesicle) or components (e.g., neurotransmitters).

This chapter will also discuss typical ANN models, which assume dendrites to be passive linear receivers of neural activity, and thus all synaptic inputs are homogeneously summed within a point-like neural unit. This total neural input is then transformed by a non-linear threshold that represents the net effect of computation in biological neurons. Therefore, ANN models also exclude dendrites and axons. The weighted connection between neural units, however, is referred to as a "synapse," despite the lack of dendrites or any other aspect of a biological synapses besides its strength. Another difference between real neurons and ANNs is that while artificial neural units can have positive and negative weights onto their targets, biological neurons can only make one type of connection to downstream cells, either positive (excitatory neurons) or negative (inhibitory neurons). This division of connection valence according to cell type is known as Dale's law (Strata and Harvey 1999). Finally, it should be noted that synapses in ANN models are inspired by chemical synapses as described above. The biological brain also has electrical synapses, based on a physical connection called a gap junction, that allow neurons to communicate directly by sharing membrane voltage. While electrical synapses have also been largely ignored by ANN models, further discussion is out of scope for this chapter. Again, a comprehensive review is beyond the scope of this text, and we will instead focus at a more abstract level.

A neuron can be modeled based on its electrical activity, arising from ion flow in and out of the neuronal cell membrane, which leads to a voltage potential difference that might lead to the generation of an action potential (spike) that will travel the length of the axon. The leaky integrate-and-fire (LIF) model is one of many ways to model a neuron (Dayan and Abbott 2001; Miller 2018), a relatively simple one that is nevertheless capable of producing emergent behaviors in the case studies below, while still remaining computationally tractable for larger network sizes. The "leaky-integration" portion of the LIF models the membrane potential V via dynamics

$$C\frac{dV(t)}{dt} = I(t) - \frac{V(t)}{R} \tag{8.1}$$

where C is the membrane capacitance, I is the neuron's input, and R is the membrane resistance. The LIF model accumulates the input signal, and when V reaches some threshold potential V_t it "fires," sending a spike of current to the downstream neurons (see next section) and resetting V to zero.

The LIF model can be generalized in a number of ways to add additional biological fidelity to neuronal dynamics. One common extension is to replace the linear integration of the input I with a non-linear term such as a quadratic or exponential rule. These non-linearities serve to essentially change the firing threshold based on the "shape" of the input $I(t)$, for example so that "fast" inputs trigger firing where slower inputs with similar area do not. Additionally, these non-linear models can be modified to introduce refractory periods, where the accumulation rate is dependent on the time since the last firing event. This latter behavior fundamentally limits the overall output rate of the neuron, potentially introducing stability to the overall network. Additional modifications for adaptation can be introduced to capture firing-rate patterns observed in nature, such as initial bursts followed by limited activity, increasing delays between spikes, delayed/transient responses, etc. We also note that the original LIF model and the extensions above have been introduced in terms of deterministic behaviors, and there are a number of mechanisms to introduce randomness to these models. One such simple mechanism is to have the firing and reset behavior of the LIF model to be based on some random firing process with the probability of firing proportional to the membrane potential V. In the context of more general emergent phenomena, the introduction of this sort of randomness introduced variability into the system that may prevent degeneracies in behavior. In summary, computational neuroscience offers a variety of modifications to the simple neuronal dynamics of Equation 8.1 that adapt or regulate the behavior of an individual neuron that could have a profound impact on the collective behavior of the network.

While the analysis below will primarily focus on the basic LIF model of a neuron as described above, we now briefly discuss an alternative class of models of neuronal dynamics, so-called rate-based models. These model a spike firing rate $v(t)$ (rather than generating spikes) and have dynamics of the form

$$\frac{dv(t)}{dt} = -v(t) + F\big(I(t)\big) \tag{8.2}$$

where F is some non-linearity, and I is again the neuron's input. The fixed points of these systems correspond to the condition $v(t) = F\big(I(t)\big)$, which should be immediately familiar to machine learning practitioners who deal with ANNs as the input-output relationship of an artificial neuron. As was the case with the LIF neuron, there are many variants (in the form of different non-linearities F) that can capture different firing patterns observed in biological neurons.

8.2.2 Neural Networks

Biological neurons can communicate with electro-chemical signals via chemical synapses. In the context of a simplified neuron model described above, synapses are generally modeled as a set of "weights" w_{ij} between neurons. If V_i denotes the LIF model for the ith neuron in a network of N neurons, then we can adapt the above LIF dynamics to account for the network structure

$$C\frac{dV_i(t)}{dt} = I(t) + \sum_{j=1}^{N} w_{ij}S_j(t) - \frac{V_i(t)}{R} \tag{8.3}$$

where $S_j(t)$ is the impulsive "spike-train" output of neuron j which causes the voltage V_i to instantaneously change by w_{ij} when neuron j fires. In reality, a synapse will introduce additional noise, non-linearity, transport delay, etc., not captured by this linear relationship. Firing-rate models are adapted similarly, with the combination of exogenous input and network inputs occurring inside a non-linearity F.

At this level of abstraction, one can analyze a neural network using the branch of mathematics known as graph theory, which we discuss in the following section. Models of network processes that use graphs are often modeled as symmetric $w_{ij} = w_{ji}$, however, such a model is not considered biologically plausible as synapses are naturally directional (requiring a directed graph model). As noted above, Dale's law states that a given neuron is either *excitatory*, meaning its spiking should produce an increase in spiking on its post-synaptic neighbors, or *inhibitory*, which tends to suppress firing in its post-synaptic neighbors. From the perspective of our network model, this implies that our network contains both negative and positive weights, corresponding to inhibitory and excitatory neurons, again violating a common approach to modeling networks as consisting entirely of non-negative weights. Furthermore, as this is a property of the neuron and not the synapse, for a given neuron j, w_{ij} must all be either non-positive (inhibitory) or non-negative (excitatory), inducing additional structure on our network model.

Taking a step back from the precise mathematical model described above, one can imagine the potential interactions between neurons (or indeed, agents in other SoS emergence scenarios) might be broadly impacted by the notion of excitatory and inhibitory interactions. Excitatory interactions result in increased activity in the network, whereas inhibitory interactions result in decreased activity in the network. These general concepts and the presence of non-linearities in the models suggest that there will likely be a delicate balance between the excitatory and inhibitory neurons that can generate the desired "positive" emergence in a neural network (i.e., the neural activity that supports its function). Any imbalances might lead to a lack of neuronal activity (due to too much inhibition) or saturation of neuronal activity (due to too little inhibition).

8.3 GRAPH SIGNAL PROCESSING

Over the past decade, the field of GSP has expanded across multiple applications and novel techniques have enabled analyses otherwise not possible. In this section, we provide the foundations for graph theory needed for GSP, followed by an overview of GSP, and will conclude with a review of graph spectral analysis and GSP in neuroscience.

8.3.1 INTRODUCTION TO GRAPH THEORY

First, we define a (undirected) graph $G = (V, E)$, where V is a set of vertices or nodes and $E = \{\{u, v\} : u, v \in V\}$ is a set of edges representing relationships among the vertices. Note that a graph is called simple if there are no self-loops ($\{u, u\} \notin E$ for all $u \in V$) and multiple edges do not exist between a single pair of vertices. For an edge $\{u, v\} \in E$, we say that u and v are adjacent and that v is a neighbor of u. Both vertices and edges may have attributes associated with them. For instance, a vertex attribute in a co-authorship graph may be a label (e.g., name of the author represented by the vertex) or a numerical value (e.g., how long the author has been publishing, how many papers they have published, etc.). An edge attribute describes the relationship. The most common edge attribute is a weight indicating the strength of the relationship. In the co-authorship network example, each edge might be weighted by the number of papers co-authored by two people. The degree of a vertex is the number of neighbors of the vertex. For $v \in V$, $deg(v) = \left|\{u : \{u, v\} \in E\}\right|$.

The adjacency matrix of a graph, indicated by A, represents the edge relationships among the vertices with $A_{ij} = 1$ if $\{v_i, v\_j\} \in E$ and $A_{ij} = 0$ otherwise. For a weighted graph, $A_{ij} = w_{ij}$ where w_{ij} is the weight of the edge $\{v_i, v_j\}$. Note that if G is simple, then $A_{ii} = 0$ for all i. Additionally, A is symmetric when G is undirected. For a simple undirected graph, the Laplacian matrix is defined as $L = D - A$, where D is a diagonal matrix with $D_{ii} = deg(v_i)$, the degree of vertex v_i and A is the adjacency matrix. The Laplacian matrix is a rich representation of a graph, as it encodes several interesting properties of the graph such as the number of spanning trees, the number of connected components, and the overall strength of connectedness of the graph.

For undirected graphs with nonnegative, real-valued weights, both the adjacency and Laplacian matrices are real and symmetric, meaning they have n (not necessarily unique) real eigenvalues and one can find a corresponding set of n orthonormal eigenvectors, $L = U\Lambda U^T$. While the eigenvalues of the adjacency matrix sum to 0, the Laplacian matrix is positive semi-definite and thus has non-negative eigenvalues.

A graph is directed if the edges have directionality. We use parentheses to indicate a directed edge so the edge set is $E = \{(u,v),\ u,v \in V\}$. Then, u is referred to at the head and v is the tail. In-degree is the number of edges coming into a vertex, denoted $d^-(v) = |\{u : (u,v) \in E\}|$. Similarly, out-degree is the number of edges leaving a vertex, denoted $d^+(v) = |\{v : (u,v) \in E\}|$. Moreover, a signed directed graph is a graph in which its edges e_{ij} can take both positive and negative values. In addition to biological neural networks, signed directed graphs are observed in multiple applications, such as social networks.

8.3.2 Introduction to GSP

The main motivation of GSP is to analyze signals over a graph, where signals now live on an irregular domain (Shuman et al. 2013). A graph signal $x \in R^N$ is defined over a graph G (with N vertices) and the nodes of the graph form the domain of the signal. An example graph with corresponding signal is depicted in Figure 8.1. The field of GSP has focused on extending techniques from classical signal processing into signals defined over graphs. Note that now the emphasis is no longer on time, but

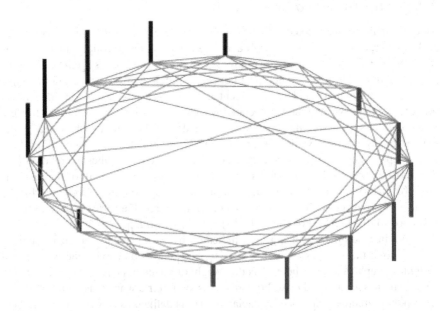

FIGURE 8.1 Illustration of graph signals over a Watts-Strogatz network. The thin lines are the edges between the nodes, and the graph signal is depicted by the thick vertical lines protruding from the top of the node for positive signal components or the bottom for negative signal components

on how the signal varies over the nodes of a graph. An important transform applied to graph signals is the graph Fourier transform (GFT), generally defined as

$$\hat{x} = U^T x, \tag{8.4}$$

where $x \in \mathbb{R}^N$ is the graph signal and $U \in \mathbb{R}^N \times \mathbb{R}^N$ is a matrix whose columns define the graph Fourier basis functions, or harmonics. Typically, these are the eigenvectors of the graph Laplacian, but the eigenvectors of the adjacency matrix can also be used as basis functions. In this chapter, the GFT of x, $\hat{x}$ is defined over the eigenvalues of L, $\lambda_0 \leq \lambda_1 \cdots \leq \lambda_{N-1}$. Now the eigenvalues of the graph Laplacian define the frequencies of the signal. For undirected, unsigned graphs, the first eigenvector u_0, corresponding to the smallest eigenvalue, is constant, an analog to a zero frequency component in classical signal processing. Eigenvectors corresponding to larger eigenvalues oscillate faster. This is a convenient property of the Laplacian spectrum for graph frequencies, which follows analog definitions as in classical Fourier analysis.

In multiple applications, the graph signals also vary over time, defining a new signal $X \in \mathbb{R}^{N \times T}$, i.e., the N nodes are discretely sampled at T time points. To analyze graph signals that vary over the graph and time, the joint vertex-time (JVT) transform extends the GFT (Grassi et al. 2017)

$$\hat{X}(l,k) = \frac{1}{\sqrt{T}} \sum_{n=1}^{N} \sum_{t=1}^{T} X_{n,t} u_{l,n}^* e^{-jw_k t}, \tag{8.5}$$

where u_l is the lth eigenvector of the graph Laplacian and $e^{-jw_k t}$ is the Fourier basis. Alternatively, in matrix form

$$\hat{X} = JVT\{X\} = U_G^\dagger X U_F^T \tag{8.6}$$

where U_G consists of the graph Laplacian eigenvectors, and U_F is the DFT matrix of appropriate dimension. The JVT essentially computes the classical Fourier transform over time of the GFT over the graph nodes.

Unlike the graphs used in many GSP applications, in biological neural networks, the graph model is both directed (due to the one-way directionality of the synapses) and signed (due to the presence of both excitatory and inhibitory neurons). From a spectral graph theory perspective, this presents several challenges as the graph Laplacian is no longer diagonalizable into an orthonormal basis for the transforms. These spectral conditions then impact the interpretation and intuition derived from the GFT. For directed graphs, multiple techniques have been proposed to obtain an orthonormal basis from the graph Laplacian or the adjacency matrix. Those methods involve either optimization approaches that impose certain constraints to derive the basis function, or approaches that propose novel representations of the graph into matrices that can produce an orthonormal basis, such as the Hermitian Laplacian Furutani et al. (2019). Here we employ the signed Hermitian Laplacian

$$L_q = D - \Gamma_q \odot A^{(s)}, \tag{8.7}$$

where D is the degree matrix of a symmetrized graph, $\Gamma_{q_{ij}} = e^{j2\pi q\left(w_{ij} - w_{ji}\right)}$, $\odot$ denotes elementwise multiplication, and $A^{(s)}$ is the symmetrized adjacency matrix, i.e., $A_{ij}^{(s)} = \frac{1}{2}\left(w_{ij} + w_{ji}\right)$.

8.3.3 GSP IN NEUROSCIENCE

In recent years, network neuroscience has emerged as a powerful tool for the study of neuronal networks across multiple scales. In particular, graph spectral analysis has been used for graph comparison, graph embeddings, and structure-function analysis. One illustration of the use of graph spectral analysis for network comparison is the work of de Lange, de Reus, and Van Den Heuvel (2014) where the connectomes from the neuronal networks of the macaque, cat, and *C. elegans* were compared to model and empirical networks by using a similarity metric based on the spectral distance from a smoothed eigenvalue distribution from the normalized Laplacian. In another work, Raj et al. (2020) developed a spectral graph model based on the spectrum of the Laplacian from the structural connectome to derive a closed-form solution to the structure-function problem. The graph spectra has been also employed in the construction of spectral graph embeddings to determine the importance of cells in *C. elegans* (Petrovic et al. 2019). Finally, Aqil et al. (2021) developed a spatiotemporal framework of dynamical models in the human connectome based on the eigenvectors of the human connectome Laplacian to study structure-function relationships.

Earlier work in GSP for neuroscience applications focused mostly on macro-scale neuroscience. Various works focused on the study of alignment Medaglia et al. (2018) and the introduction of GSP for neuroimaging data (Huang et al. 2018; Goldsberry et al. 2017). GSP wavelets have also provided significant contributions to the study of macro-scale connectivity in the human brain (Leonardi and Van De Ville 2013), including novel ways to construct connectomes (Behjat et al. 2015). Other works focused on the study of fMRI temporal analysis using GSP techniques (Brahim and Farrugia 2020).

8.4 EMERGENCE IN THE SHORT TERM: COLLECTIVE ACTIVITY

Collective activity in neural networks can take place in many forms, including activity bumps, oscillations, and traveling waves. Activity bumps are when a subset of the neurons fire in response to an external (from the perspective of a given neural network) input signal. This type of behavior should be familiar to machine learning practitioners where, for example, inputs to convolutional neural networks produce feature extraction as activity bumps across the hidden layers (Lindsay 2021). Unsurprisingly, since these networks and related structures derive motivation from the visual cortex, activity bumps are prevalent there and other sensory receptive fields. Some forms of activity bumps maintain their activity even if the external input is removed, resulting in so-called self-sustaining activity bumps. This form of collective behavior is especially prevalent in the hippocampal formation where spatial computation and estimation is performed in the form of place, head direction, and grid cells, among others (Knierim and Zhang 2012). A leading theory of these stable activities is *attractor theory* where the emergent bumps are controlled by these external inputs, yet the bumps remain active even when these inputs are removed due to "attraction" of the dynamics to the low energy states of the network. This theory was originally formulated as a mechanism for memory (Hopfield 1982; Amit and Treves 1989) that could store discrete patterns as fixed points of the network dynamics. However, as space is itself continuous, attractor theory was soon extended to so-called continuous attractors (Samsonovich and McNaughton 1997), where instead of stable fixed points, the dynamics of the network are attracted to stable sub-manifolds of equal energy.

The above emergent phenomena are analogous to collective behaviors in space, since the collective firing is occurring roughly simultaneously, in response to some external input (or persistently, in the case of self-sustaining activity bumps). There are also examples of emergent collective firing that are more time-oriented, such as oscillations. There are numerous emergent oscillations in the brain ranging from 0.02 to 600 Hz (Penttonen and Buzsáki 2003). These oscillations are important because they form a hierarchical framework for action potentials to traffic within and across neuronal circuits at many temporal scales (Buzsáki 2019; Monaco, Rajan, and Hwang 2021). When viewed together, these oscillation bands form a linear progression on a natural log scale, spanning ten frequency bands. These frequency bands can co-exist and interact with each other in the brain in the same or different structures giving rise to various brain states (e.g., task engagement or sleep). Many of these frequency bands are thought to be nested in which the phase of the slower oscillation modulates the amplitude of a faster oscillation, and in turn that phase of the faster oscillation modulates the amplitude of the even faster oscillation and so on. Many of these frequency bands have been observed across many species, and some have been given names. In rodent studies, entrainment of theta oscillations (4–10 Hz) is required to enable movement (Fuhrmann et al. 2015), while the frequency of theta oscillations modulate movement speed (McNaughton, Barnes, and O'Keefe 1983). In contrast, sharp wave ripples (100–200 Hz) are known to occur transiently during deliberate moments of immobility (Pfeiffer and Foster 2013) in spatial memory tasks. Monaco et al. (2021) proposed that these hierarchically nested oscillations are reentrant flows on recurrent networks that can form a new computational basis. Sharp wave ripples are an example of a traveling wave (that happens to be oscillatory), where a set of neurons fire in sequence.

In the following sections, we consider a pair of simulation examples from the computational neuroscience literature to illustrate the utility of GSP in understanding the collective behaviors of the neurons. These examples focus on the analysis of activity bumps as these are more analogous to the spatial motivation of GSP, as opposed to purely oscillatory behaviors that are readily identified by standard time-domain (Fourier) analysis. That said, the combination of GSP and standard Fourier analysis is a powerful tool for understanding joint spatial and temporal dynamics. To this end, in the latter example, we consider a network that is driven by an oscillatory input, resulting in oscillatory outputs that give a sense of the potential for the JVT analysis of neural networks. This behavior could also be viewed as a stationary traveling wave, and as such serves as an example for the ability of the JVT to reveal joint spatiotemporal structure in stationary emergent behaviors. Transient traveling waves are by their very nature non-stationary and would require additional tools form non-stationary signal processing such as (graph) wavelets, which is beyond the scope of this chapter.

8.4.1 CASE STUDY: SLOW SWITCHING ASSEMBLIES

Schaub et al. (2015) used a basic LIF model of a neuron along with a series of structured network models to produce coordinated firing whose behavior is interpretable with the network structure. The first model we will consider from Schaub et al. (2015) consists of densely connected "blocks" of excitatory neurons with only sparse connections between these densely connected blocks. Unlike the excitatory neurons, the inhibitory neurons are connected uniformly at random, as are the connections between excitatory and inhibitory (and vice versa). An example network is shown in Figure 8.2a, which was actually generated using the learning rules discussed in the following section. In this particular instance, there are 100 excitatory neurons split into six densely connected blocks and 25 inhibitory neurons.

As found in Schaub et al. (2015), when the excitatory neurons are stimulated by a critical level of external stimuli, these blocks exhibit bursts of activity within a densely connected block that slowly (and chaotically) transitions between blocks. An example of this slow switching assembly (SSA) behavior is shown in Figure 8.2b, illustrating that activity is concentrated within one block at a time, and exhibits slow, random switches between blocks. The inhibitory neurons are essentially randomly activated throughout the process, yet must play a key role in regulating the overall emergent behavior of the neural network. In particular, Schaub et al. (2015) note

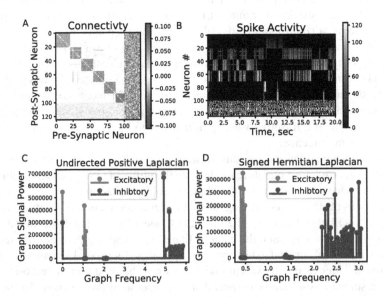

FIGURE 8.2 Slow-switching assembly simulations. (Adapted from Schaub et al. 2015.) (a) Adjacency matrix showing block structure. (b) Neuronal activity showing slow switching behavior between blocks. (c) GFT power using the undirected positive Laplacian. (d) GFT power using the signed Hermitian Laplacian.

that the inputs to the inhibitory networks must be slightly larger than the excitatory weights in order to maintain stability of the network.

In Schaub et al. (2015), analysis of this collective activity was considered in the context of the Schur decomposition of the network connectivity matrix. This linear algebraic technique could, in theory, be used to perform GSP analysis, since it produces a set of orthonormal basis vectors. However, the Schur decomposition is not unique, leading to a lack of reproducibility. As discussed above, since this network is both signed and directed, we need to consider GFTs that accommodate this additional structure. The first transform we consider uses the underlying undirected transform, and the graph signal power with respect to this transform is shown in Figure 8.2c. This transform captures most of the signal power of the excitatory network into a few harmonics, but there are more contributing harmonics than there are blocks in the network model. Additionally, the inhibitory and excitatory portions of the network are not totally separated, partially due to the constant harmonic, but also in the higher frequency harmonics.

In contrast to the underlying undirected transform, accounting for the signed and directed nature of the network model produces a graph Fourier power spectrum that clearly separates the excitatory and inhibitory portions of the network (see Figure 8.2d). Furthermore, the number of harmonics that capture the excitatory signal power is equal to the number of blocks in the excitatory portion of the network. The inhibitory portion of the network, on the other hand, is reasonably evenly split in power among 25 harmonics, consistent with the notion that they are seemingly firing at random. We will further explore the distribution of harmonics in this network in Section 5.1.

8.4.2 CASE STUDY: FRUIT FLY PROTOCEREBRAL BRIDGE

Another form of emergent collective firing in neural networks is self-sustained activity bumps, that is, localized regions of increased activity in a contiguous portion of the network. The output activity of the network may be guided by feed-forward inputs to the network that can manipulate the activity bump, but in the absence of such inputs the activity bump will be maintained. This sort of emergent behavior is prominent in neuronal circuits associated with navigation, where the inference of position and orientation must be maintained even in the absence of stimuli. Additionally, these circuits should be robust to noise such as random firings of both neurons within the network, and those modeled by the feedforward inputs. As such, seemingly chaotic coordinated firing as observed in the previous section represents a fundamentally different emergent behavior. To study this phenomenon through the lens of GSP, we use a simulation of the fruit fly protocerebral bridge (Kakaria and de Bivort 2017), which is believed to be responsible for an egocentric estimate of the fly's heading.

Unlike the SSA model from the previous example, the model of the fruit fly protocerebral bridge is structured in a fundamentally different way than the stochastic block model above (see Figure 8.3a). Here, the excitatory portion of the network results in ring-attractor dynamics with a corresponding triple of interlocking ring graphs (for full anatomic details see Kakaria and de Bivort (2017)). These ring attractor dynamics are exploited to maintain the heading estimate of the fruit fly. Additionally, unlike

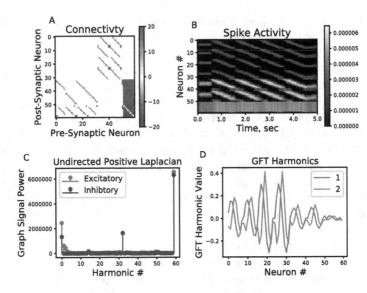

FIGURE 8.3 Fruit fly protocerebral bridge simulations. (Adapted from Kakaria and de Bivort, 2017.) (a) Adjacency matrix for the model. (b) Neural activity to a fixed angular velocity stimulus applied at 0.5 second for 4 seconds. (c) GFT power of the neural outputs. (d) GFT harmonics of the non-trivial active harmonics that are approximately sinusoidal and 90° out of phase on the excitatory neurons.

the SSA model, the weights of the synapses are chosen from similar, discrete values, and not the more random weights as above. The activity of the network is shown in Figure 8.3b. Unlike the SSA example above, this model expects a more structured feed-forward input that corresponds to the output of unmodeled upstream neurons. In the absence of this input signal, the network maintains a fixed activity bump corresponding to the last state estimate. Here, this corresponds to the times before 0.5 second and after 4.5 seconds. In the intervening times, a stimulus is applied that corresponds to a 1 Hz rotation, which causes the activity bumps on the excitatory neurons to rotate at that rate.

Using the undirected, positive Laplacian as the basis for a GFT we see considerable structure in the graph power spectrum (see Figure 8.3b). Harmonics 0, 32, and 59 capture the average contributions between the major functional components at each point in time. The next two contributing harmonics (1 and 2) are shown in Figure 8.3d. These harmonics resemble a pair of sinusoidal waves on the excitatory networks that are 90° out of phase. Such harmonics are characteristic of ring-like networks, and these harmonics contribute to the localized activity bumps that encode the heading of the fruit fly. In contrast, using the signed Hermitian Laplacian (not shown) appears to group the network into three regimes (0–31, 32–49, 50–59), and only has the ring-like harmonics on neurons (0–31).

Next, we further explore the interplay between GSP and emergence in neural networks by highlighting the ability of GSP to identify hidden patterns of coordinated activity in time and space. When the neurons are enumerated as in Figure 8.3b, the coordinated firing activity is quite obvious, although one might question why the activity bump is repeated six times (or twice on each functional component). However, if the natural ordering were not known a priori, and instead we are presented with a random permutation of the neuronal indices, then the output is considerably harder to decipher (see Figure 8.4a). One might note that many of the neurons appear to have some periodic behavior, and standard Fourier analysis of the individual neurons indicates that many of the neurons do indeed have considerable power contributions at 1 Hz.

An important feature of the GFT is that it is "invariant" to permutations of the vertices, in the sense that individual harmonic vertex values will be permuted in the same way. Thus, the GFT power spectrum using the underlying undirected positive Laplacian is identical to that in Figure 8.3c. Armed with the knowledge of the structure from Fourier analysis in the time and vertex domains individually, we next considered the JVT transform (Figure 8.4c). This reveals that the 1 Hz power observed in the individual neurons in Figure 8.4b is strongly concentrated in just two harmonics, and these are of course harmonics 1 and 2. With the permutation of the neurons, these harmonics do not have the same readily apparent structure as in Figure 8.3d. Since these two harmonics have nearly identical graph frequency, one might be tempted to think of them in an analogous manner to the real and imaginary parts of a standard Fourier complex exponential harmonic. With this intuition, re-permuting the neuron index by the "phase" of the combined harmonic $U_1 + jU_2$ at

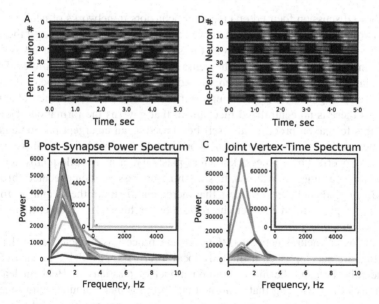

FIGURE 8.4 Permuting the index of the protocerebral bridge simulations. (a) Neuronal outputs when the neuron indices are permuted (i.e., the rows of Fig. 3b are permuted – the data is itself identical). (b) Power spectrum of neuronal outputs, illustrating concentration around 1 Hz. (c) JVT power spectrum, illustrating concentration of 1 Hz power in to two harmonics (note scale difference from b). (d) Re-permuting the neuron index using the phase between each neuron's contribution to the dominant JVT harmonic, clearly revealing rotational input.

each neuron yields Figure 8.4d. This sorting completely unravels both the permutation and the original interleaved structure and shows how the activity bump travels smoothly across the excitatory portion of the network.

8.4.3 SUMMARY

In summary, neural networks exhibit many different forms of emergent collective activity that is coordinated spatially, temporally, or both. The synaptic network between the neurons is key in producing these collective behaviors, with the connectivity and weights ultimately responsible for generating the collective dynamics. Commonalities between the examples' network structure reveal a motif for potential consideration for wider SoS scenarios. This motif is that of structured, yet sparse, connectivity in the excitatory portion that demonstrates the "core" of some emergent behavior, coupled with dense connectivity involving a smaller inhibitory portion that serves to regulate the overall collective behavior. Additionally, we showed that GSP can be used to reveal low-dimensional hidden structure in time and space for these collective behaviors, but transforms that account for the directed nature and presence of inhibitory interactions may be required to fully reveal this structure.

8.5 EMERGENCE IN THE LONG TERM: NETWORK FORMATION AND LEARNING

The above discussion focused on emergent behaviors in networks of neurons where the network between the neurons is viewed as static in both the weights of the model as well as the connectivity. However, the structure of the network itself is fundamental in the overall function of the neural network. This is especially evident for self-sustaining activity bumps when viewed through the lens of attractor theory, as the network weights are a major component in the determination of the stable manifolds. As there is no centralized mechanism that governs the formation of the network, this formation process can itself be viewed as an emergent phenomena. The formation of the connections in the network and their strength (i.e., their weight in a model) is governed by a process known as plasticity, or more colloquially, learning. In machine learning applications, this learning process is often executed through a centralized gradient descent process that incrementally tunes the weights of the network from a pre-defined connectivity pattern or "architecture" in order to minimize some loss function.

In biological neural networks, the learning process is modeled using learning rules that are accomplished using only local information available to each neuron, individually. Perhaps the most common model of plasticity is Hebbian learning, which uses the reasoning that neurons that "fire together" should "wire together," that is, coordinated firing should strengthen synaptic connectivity, and vice versa. The most common formulation of this approach is in terms of firing rate models, where the synaptic weights change via

$$\frac{dw_{ij}}{dt} = g\left(w_{ij}, v_i, v_j\right),$$

(8.8)

where g is a function of the current weight and the firing rates v_i of the neurons. The most basic form of Hebbian learning is $g(w_{ij}, v_i, v_j) = \eta_{ij} v_i v_j$, where η_{ij} is referred to as the learning rate.

While computationally tractable and intuitively appealing, this form of "pure" Hebbian learning has some shortcomings with respect to the network constraints observed in biological neural networks. First, it is clear that the learning rule above is symmetric, which will always result in undirected network models. Second, this approach to learning was originally intended for excitatory neurons only and does not account for inhibitory neurons. Finally, the dynamics of Hebbian learning often leads to instability, with exponential growth in the magnitudes of the weights. This latter failing can be addressed by various normalization techniques, leading to alternative learning rules such as Oja's $g(w_{ij}, v_i, v_j) = \eta(v_i v_j - w_{ij} v_i^2)$, which asymptotically normalizes $\sum_j w_{ij}^2$ to 1.

The above learning rules are appropriate for firing rate models, with their rates η_i defined at all times. For spike-based models with their discrete spiking events, the learning rules should be dependent on the time between the pre-synaptic spike and the post-synaptic spike, so-called spike-timing-dependent plasticity (STDP). If we define t_k^i to be the time of the k th spike in the spike train S_i and t_l^j for the train S_j, a basic form of STDP is

$$\frac{dw_{ij}}{dt} = \sum_k \sum_l h(t_k^i - t_l^j), \tag{8.9}$$

where $h(t)$ is called the learning window. A basic form of h is

$$h(t) = \begin{cases} A_+ \exp -\dfrac{t}{\tau_+}, & for\ t > 0, \\[2mm] -A_- \exp \dfrac{t}{\tau_-}, & for\ t < 0, \end{cases} \tag{8.10}$$

with $A_\pm$, $\tau_\pm$ positive, and $A_\pm$ may depend on w_{ij}. As with the rate-based Hebbian learning above, there are many variants of the STDP rules that capture different biologically observed phenomena. In particular, STDP can serve as both a model for directed connectivity as well as for a method of plasticity for inhibitory neurons.

8.5.1 Case Study: Structured Assembly Formation

In Triplett, Avitan, and Goodhill (2018), a Hebbian-like learning rule was used to evolve the excitatory portion of a neural network from an initially uniformly randomly connected network to one that exhibits both the stochastic block structure and the slow switching behavior of Schaub et al. (2015). In a sense, this is essentially a generative mechanism for the dynamics of Schaub et al. (2015), although we note that the weights of any edges connected to the inhibitory network are fixed. There, for ease of simulation, the LIF dynamics were simplified and time discretized, simulating the responses of the excitatory neurons S_i^E and inhibitory neurons S_i^I via

$$S_i^E(t+1) = \Theta\left(\sum_j w_{ij}^{EE} S_j^E(t) - \sum_j w_{ij}^{EI} S_j^I(t) + \beta_i^E(t) - \gamma\right)$$

(8.11)

$$S_i^I(t+1) = \Theta\left(\sum_j w_{ij}^{IE} S_j^E(t) - \sum_j w_{ij}^{II} S_j^I(t) + \beta_i^I(t) - \gamma\right),$$

where w_{ij}^{EE}, w_{ij}^{EI}, w_{ij}^{IE}, and w_{ij}^{II} are the excitatory-to-excitatory, inhibitory-to-excitatory, excitatory-to-inhibitory, and inhibitory-to-inhibitory portions of the network, respectively, Θ is the Heaviside step function, γ is the activation threshold, and $\beta_i^E(t)$, $\beta_i^I(t)$ are random variables that drive spontaneous background activity. The model of Triplett, Avitan, and Goodhill (2018) uses a covariance learning rule, where the weights of the excitatory sub-network w_{ij}^{EE} are updated via

$$\Delta w_{ij}^{EE}(t) = \eta\left(S_i^E(t) - S_i^E(t)\right)\left(S_j^E(t) - S_j^E(t)\right),$$

(8.12)

where $\langle S_i^E(t)\rangle$ is the running average of $S_i^E(t)$ and η is the learning rate. Furthermore, the weight updates are constrained to prevent negative excitatory weights normalized to maintain a constant sum of weight for each excitatory neuron. This latter step is needed to prevent the "rich-get richer" phenomenon where all of the synaptic weight accumulates in a single edge.

As shown in Triplett, Avitan, and Goodhill (2018), the combination of these neuronal dynamics and learning rule results in the formation of strongly connected block models with slow-switching behavior in the vein of (Schaub et al. 2015). In fact, a simulation of this process was used to generate the network connectivity used in Section 8.1, that was then used as the connectivity for the higher-fidelity dynamics of (Schaub et al. 2015). Figure 8.5a shows how the number of strongly connected components evolves over time, increasing monotonically from the single initial component to six components. In the context of the GFT, the evolution of this network can be interpreted in terms of the associated graph frequencies, see Figure 8.5b. This figure shows that the formation of strongly connected blocks in the excitatory subnetwork are associated with the appearance of a low-frequency graph harmonic in the GFT. Of course, this result does not hold for the undirected transform, only the transform that accounts for the signed and directed nature of the graph.

8.6 CONCLUSION

In conclusion, biological neural networks exhibit many different forms of emergent coordinated firing activity that evolve both temporally and spatially across the network. Furthermore, the very process of network formation is itself an example of an emergent behavior on a much longer time scale. As a process that takes place on a natural network structure, tools from GSP can be applied to analyze these emergent

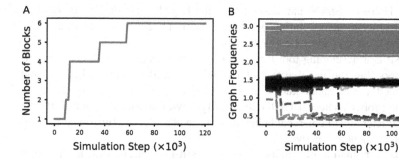

FIGURE 8.5 Learning induced block formation. (Adapted from Triplett, Avitan, and Goodhill, 2018.) (a) Number of blocks over time, eventually resulting in the network shown in Fig. 2a. (b) Evolution of graph frequencies over time, using signed Hermitian Laplacian. High-frequencies ($\lambda > 2$) cover the inhibitory neurons, whereas lower frequencies consist primarily of excitatory neurons. The dashed-colored lines are those that eventually map to a block in the model, which forms when the frequencies drop below the main group of frequencies.

behaviors, revealing insight into these processes. Given the rich history of interconnections between signal processing and control theory, this also points toward a future potential capability to engineer desired emergent behaviors. Looking beyond biological neural networks to the perspective of general SoS, biological neurons and neural networks have a number of interesting features to motivate the design of SoS, and GSP is a natural tool to consider these emergent SoS behaviors. Specific facets discussed in this chapter include:

- Collective emergent behaviors in neural networks assumes many forms beyond standard consensus and synchronization behaviors and may serve as inspiration for a number of other application areas. Many problems can presumably find neuromorphic or neuromimetic solutions once the appropriate neural circuit and conversion to problem domain is identified, see e.g., the swarming approach considered in Monaco et al. (2020).
- The structure of the neural network is key to the presence of emergence, and this structure must form organically, itself an example of emergence on a longer time scale than the shorter dynamical time scale.
- Neurons and the synapses that interconnect them have a number of regulatory processes that allow for interesting collective behaviors to emerge. These include
  - Excitatory and inhibitory neurons. In the examples studied here, what we would identify as the emergent behavior is observed primarily in the excitatory neurons, but the inhibitory neurons are needed to stabilize the behavior.
  - Auto-regulatory behaviors that vary the responsiveness of a neuron to inputs in order to prevent saturation.
  - Regulatory processes for the weakening and strengthening of synapses (network plasticity) to limit the rate of network change, maintain certain network features, and/or prevent degeneracies.

These processes and the motifs in the way they are employed could be valuable in understanding potential designs for emergence in other SoS, even those that are not explicitly neuro-inspired.

- GSP is a developing tool for analyzing dynamics and interactions on networks, but the peculiarities of biological neural networks (directed and signed) are an under-studied area.
- The proper choice of GFT can identify low-dimensional structure that is characteristic of emergent behavior and can also identify functional components in a neural network.
- The combination of graphical and standard Fourier analysis, leading to a JVT, can further reveal this low-dimensional behavior by detecting behaviors that evolve both temporally and spatially across the network.

ACKNOWLEDGMENTS

This work was supported by NSF NCS/FO 1835279 and JHUAPL IR&D funds. This material is based on work supported by (while serving at) the National Science Foundation. Any opinion, findings, and conclusions or recommendations expressed in this material are those of the authors and do not necessarily reflect the views of the National Science Foundation.

REFERENCES

Amit, Daniel J., and Alessandro Treves. 1989. "Associative memory neural network with low temporal spiking rates." *Proceedings of the National Academy of Sciences of the United States of America* 86(20): 7871–7875.

Aqil, Marco, Selen Atasoy, Morten L Kringelbach, and Rikkert Hindriks. 2021. "Graph neural fields: A framework for spatiotemporal dynamical models on the human connectome." *PLoS Computational Biology* 17(1): e1008310.

Badimon, Ana, Hayley J Strasburger, Pinar Ayata, Xinhong Chen, Aditya Nair, Ako Ikegami, Philip Hwang, et al. 2020. "Negative feedback control of neuronal activity by microglia." *Nature* 586(7829): 417–423.

Behjat, Hamid, Nora Leonardi, Leif Sörnmo, and Dimitri Van De Ville. 2015. "Anatomically-adapted graph wavelets for improved group-level fMRI activation mapping." *NeuroImage* 123: 185–199.

Brahim, Abdelbasset, and Nicolas Farrugia. 2020. "Graph Fourier transform of fMRI temporal signals based on an averaged structural connectome for the classification of neuroimaging." *Artificial Intelligence in Medicine* 106: 101870.

Buzsáki, Gyorgy. 2019. *The brain from inside out.* New York: Oxford University Press.

Chung, Fan RK. 1997. *Spectral graph theory.* American Mathematical Soc. http://dx.doi.org/10.1090/cbms/092

Dayan, Peter, and Laurence F Abbott. 2001. *Theoretical neuroscience: Computational and mathematical modeling of neural systems.* Computational Neuroscience Series. Cambridge: The MIT Press.

de Lange, Siemon, Marcel de Reus, and Martijn Van Den Heuvel. 2014. "The Laplacian spectrum of neural networks." *Frontiers in Computational Neuroscience* 7: 189.

Fuhrmann, Falko, Daniel Justus, Liudmila Sosulina, Hiroshi Kaneko, Tatjana Beutel, Detlef Friedrichs, Susanne Schoch, Martin Karl Schwarz, Martin Fuhrmann, and Stefan Remy. 2015. "Locomotion, theta oscillations, and the speed-correlated firing of hippocampal neurons are controlled by a medial septal glutamatergic circuit." *Neuron* 86(5): 1253–1264.

Furber, Steve. 2016. "Large-scale neuromorphic computing systems." *Journal of Neural Engineering* 13(5): 051001.

Furutani, Satoshi, Toshiki Shibahara, Mitsuaki Akiyama, Kunio Hato, and Masaki Aida. 2019. "Graph Signal Processing for Directed Graphs Based on the Hermitian Laplacian." In *Joint European Conference on Machine Learning and Knowledge Discovery in Databases*, 447–463.

Gidon, Albert, Timothy Adam Zolnik, Pawel Fidzinski, Felix Bolduan, Athanasia Papoutsi, Panayiota Poirazi, Martin Holtkamp, Imre Vida, and Matthew Evan Larkum. 2020. "Dendritic action potentials and computation in human layer 2/3 cortical neurons." *Science* 367(6473): 83–87.

Goldsberry, Leah, Weiyu Huang, Nicholas F Wymbs, Scott T Grafton, Danielle S Bassett, and Alejandro Ribeiro. 2017. "Brain signal analytics from graph signal processing perspective." In *2017 IEEE International Conference on Acoustics, Speech and Signal Processing (ICASSP)*, 851–855. IEEE.

Grassi, Francesco, Andreas Loukas, Nathanaël Perraudin, and Benjamin Ricaud. 2017. "A time-vertex signal processing framework: Scalable processing and meaningful representations for time-series on graphs." *IEEE Transactions on Signal Processing* 66(3): 817–829.

Hopfield, John J. 1982. "Neural networks and physical systems with emergent collective computational abilities." *Proceedings of the National Academy of Sciences of the United States of America* 79(8): 2554–2558.

Huang, Weiyu, Thomas AW Bolton, John D Medaglia, Danielle S Bassett, Alejandro Ribeiro, and Dimitri Van De Ville. 2018. "Graph signal processing of human brain imaging data." In *2018 IEEE International Conference on Acoustics, Speech and Signal Processing (ICASSP)*, 980–984. IEEE.

Kakaria, Kyobi S, and Benjamin L de Bivort. 2017. "Ring attractor dynamics emerge from a spiking model of the entire protocerebral bridge." *Frontiers in Behavioral Neuroscience* 11: 8.

Knierim, James J, and Kechen Zhang. 2012. "Attractor dynamics of spatially correlated neural activity in the limbic system." *Annual Review of Neuroscience* 35: 267–285.

Leonardi, Nora, and Dimitri Van De Ville. 2013. "Tight wavelet frames on multislice graphs." *IEEE Transactions on Signal Processing* 61(13): 3357–3367.

Lindsay, Grace W. 2021. "Convolutional neural networks as a model of the visual system: Past, present, and future." *Journal of Cognitive Neuroscience* 33(10): 2017–2031.

McNaughton, Bruce L, Carol A Barnes, and JJEBR O'Keefe. 1983. "The contributions of position, direction, and velocity to single unit activity in the hippocampus of freely-moving rats." *Experimental Brain Research* 52(1): 41–49.

Medaglia, John D, Weiyu Huang, Elisabeth A Karuza, Apoorva Kelkar, Sharon L Thompson-Schill, Alejandro Ribeiro, and Danielle S Bassett. 2018. "Functional alignment with anatomical networks is associated with cognitive flexibility." *Nature Human Behaviour* 2(2): 156–164.

Miller, Paul. 2018. *An introductory course in computational neuroscience.* Cambridge: The MIT Press.

Monaco, Joseph D, Grace M Hwang, Kevin M Schultz, and Kechen Zhang. 2020. "Cognitive swarming in complex environments with attractor dynamics and oscillatory computing." *Biological Cybernetics* 114(2): 269–284.

Monaco, Joseph D, Kanaka Rajan, and Grace M Hwang. 2021. "A brain basis of dynamical intelligence for AI and computational neuroscience." *arXiv preprint arXiv:2105.07284.*

Penttonen, Markku, and György Buzsáki. 2003. "Natural logarithmic relationship between brain oscillators." *Thalamus & Related Systems* 2(2): 145–152.

Petrovic, Miljan, Thomas AW Bolton, Maria Giulia Preti, Raphaël Liégeois, and Dimitri Van De Ville. 2019. "Guided graph spectral embedding: Application to the C. elegans connectome." *Network Neuroscience* 3(3): 807–826.

Pfeiffer, Brad E, and David J Foster. 2013. "Hippocampal place-cell sequences depict future paths to remembered goals." *Nature* 497(7447): 74–79.

Poirazi, Panayiota, and Athanasia Papoutsi. 2020. "Illuminating dendritic function with computational models." *Nature Reviews Neuroscience* 21(6): 303–321.

Poulter, Steven, Tom Hartley, and Colin Lever. 2018. "The neurobiology of mammalian navigation." *Current Biology* 28(17): R1023–R1042.

Puschel, Markus, and José MF Moura. 2008. "Algebraic signal processing theory: Foundation and 1-D time." *IEEE Transactions on Signal Processing* 56(8): 3572–3585.

Raj, Ashish, Chang Cai, Xihe Xie, Eva Palacios, Julia Owen, Pratik Mukherjee, and Srikantan Nagarajan. 2020. "Spectral graph theory of brain oscillations." *Human Brain Mapping* 41(11): 2980–2998.

Samsonovich, Alexei, and Bruce L McNaughton. 1997. "Path integration and cognitive mapping in a continuous attractor neural network model." *Journal of Neuroscience* 17(15): 5900–5920.

Sandryhaila, Aliaksei, and José MF Moura. 2013. "Discrete signal processing on graphs." *IEEE Transactions on Signal Processing* 61(7): 1644–1656.

Schaub, Michael T, Yazan N Billeh, Costas A Anastassiou, Christof Koch, and Mauricio Barahona. 2015. "Emergence of slow-switching assemblies in structured neuronal networks." *PLoS Computational Biology* 11(7): e1004196.

Schmidhuber, Jürgen. 2015. "Deep learning in neural networks: An overview." *Neural Networks* 61: 85–117.

Shuman, David, Sunil Narang, Pascal Frossard, Antonio Ortega, and Pierre Vandergheynst. 2013. "The emerging field of signal processing on graphs: Extending high-dimensional data analysis to networks and other irregular domains." *IEEE Signal Processing Magazine* 3(30): 83–98.

Strata, Piergiorgio, and Robin Harvey. 1999. "Dale's principle." *Brain Research Bulletin* 50(5): 349–350.

Tanner, Herbert G, Ali Jadbabaie, and George J Pappas. 2007. "Flocking in fixed and switching networks." *IEEE Transactions on Automatic Control* 52(5): 863–868.

Trianni, Vito, Elio Tuci, Kevin M Passino, and James AR Marshall. 2011. "Swarm cognition: An interdisciplinary approach to the study of self-organising biological collectives." *Swarm Intelligence* 5(1): 3–18.

Triplett, Marcus A, Lilach Avitan, and Geoffrey J Goodhill. 2018. "Emergence of spontaneous assembly activity in developing neural networks without afferent input." *PLoS Computational Biology* 14(9): e1006421.

9 Deepwater Horizon
Emergent Behavior in a System of Systems Disaster

Polinpapilinho F. Katina
University of South Carolina Upstate

Charles B. Keating
Old Dominion University

CONTENTS

DOI: 10.1201/9781003160816-11

9.1 INTRODUCTION

Emergence is a critical challenge to the escalating complexity being experienced with modern systems. Emergence is a classical systems principle impacting effectiveness in system of systems (SoS) solutions. At a most fundamental level, emergence holds that as a complex SoS operates, there are properties (behavior, structure, performance) that only come about and are recognized as the SoS operates. In short, emergence holds that patterns/properties in a complex system will come about (emerge) through operation of the system. These 'emergent' properties cannot be known in advance, predicted, or attributed to individual constituent systems/components comprising the SoS. Instead, these properties only come forth (emerge) as the individual constituent systems interact with one another and the environment. Thus, the focus for emergence is on the interactions between elements in an SoS, or between systems in an SoS. The interaction is what produces an emergent property, not the characteristics of the constituent systems. Take away the interaction and the emergent property is lost. These emergent patterns/properties cannot be anticipated beforehand and are not capable of being deduced from the understanding of system constituents or their individual properties. However, the residual effects, positive or negative, set in motion from the emergence do not retreat when the interaction ceases. This simple concept has broad ramifications for understanding emergence and working toward better practices to 'tame' this unwieldly phenomenon.Emergence has always existed in the engineering of complex system solutions, most commonly known as the 'law of unintended consequences'. This catchall concept is somehow reassuring that we are not held hostage to notions of pure mysticism our attempts to understand behaviors that appear to be unexplainable. However, for SoS, emergence has increased significance for two primary reasons. First, the scope of an SoS (e.g., cyberphysical security system) is beyond a single system focus for solution. The complex SoS requires the integration of many, potentially disparate, systems to perform in unison as an SoS. This entails being capable of dealing with the inevitable emergence which will be experienced by an SoS. Second, the design, execution, and development of an SoS is 'holistic', with emergence occurring across the range of technical, human/social, managerial/organizational, policy, and political dimensions that are characteristic of the SoS.

To explore emergence in complex SoS, an appreciation of the forces driving modern complex systems is an important consideration. Emergence as a 'fact of life' for modern systems bears amplification. There are forces that create the conditions under which emergence can manifest itself in various forms. To summarize these forces, we capture the essence from previous works. There are a multitude of increasing pressures stemming from complex systems and their inevitable problems. The cadence of this reality for practitioners might be captured in six themes, presented as forces, that permeate the modern landscape of complex systems. These themes have been extolled in various forms in numerous prior works (Jaradat et al., 2014; Keating and Katina 2011; 2012; Keating, 2014; Keating and Katina, 2019). Following these works, we summarize the forces and their implications for emergence development in Table 9.1. Interestingly, although these conditions are not 'new' in the sense that they just arrived, it seems as though success in dealing with them has had minimal effectiveness. We can also see that emergence takes a front and center position in the realization of the impacts of these forces for complex SoS. There is a myriad of approaches, both past and present, with good intentions to address our deteriorating conditions of complex systems. However, we seem to be continually confounded with abysmal results, high human costs, and an increasing array of approaches and advice that falls short of expectations. It would be shortsighted to discount the implications of this set of forces in examination of, and practices related to, dealing more effectively with emergence.

TABLE 9.1

Forces Impacting Complex Systems and Implications for Emergence

Forces	Explanation	Implications for Emergence Development
Uncertainty Escalation	At a most basic level, uncertainty suggests that precise cause-effect relationships cannot be known for complex systems. Thus, normal approaches that assume complete knowledge and deterministic analysis (mathematical formulations) to address complex systems are incompatible with systems marked by high levels of uncertainty. As complex systems become more 'complex', uncertainty will rise – as will the inability of traditional reductionist-based approaches to successfully resolve issues.	System development must take into consideration that uncertainty will pose several challenges, including: (1) inevitable fallibility of any approach, which requires constant questioning and adjustment; (2) appreciation of the uniqueness of each complex system, thus requiring an equally unique approach and subsequent journey for development; (3) expectations must be tempered for development outcomes, as the precise results cannot be known or predicted in advance; and (4) uncertainty is consistent with experiencing emergence in complex SoS, as the consequences of emergence cannot be precisely known in advance.
Ambiguity Propagation	Complete knowledge and understanding for complex systems are illusionary propositions. Instead, there will always be a lack of clarity concerning the nature of each unique SoS, the unique domain within which it exists, and the unique context within which the system is embedded. Knowledge and understanding are continually refocused as they emerge over time, and new knowledge of a complex SoS continually unfolds. The result is high levels of ambiguity or a lack of clarity. This lack of clarity is not necessarily due to carelessness, omission, or intentional ignorance. Instead, it is natural and should be expected for complex SoS.	Ambiguity is simply a product of incomplete understanding of complex SoS and their context. Irrespective of desire or intent, complex SoS will always exist in conditions of incomplete understanding. Therefore, their development will also be mired in conditions of incomplete understanding. The challenge for SoS development is found in the necessity for continuous accounting for ambiguity. This entails reduction where possible, acceptance where necessary, and accounting for development influences where feasible. Ambiguity is a byproduct of emergence in SoS.

(Continued)

TABLE 9.1 (*Continued*)
Forces Impacting Complex Systems and Implications for Emergence

Forces	Explanation	Implications for Emergence Development
Complexity Acceleration	Central characteristics of complexity include a large number of richly interrelated elements, dynamic shifting of the SoS and our knowledge of that SoS over time, and emergence of unpredictable behavior, structure, and performance over time and operation of the system. For complex SoS, complexity is not a temporary condition that will de-escalate over time. On the contrary, complexity, and its inevitable impacts, will continue to escalate as SoS evolves and becomes increasingly interconnected, unknown, and unpredictable.	SoS development must accept that complexity is not going to diminish and will most likely be exacerbated by prevailing trends. Implications for system development include finding better ways of dealing with elaboration (increasing interconnectedness), emergence (unpredictable patterns), and dynamics (rapid changes). Purposeful SoS development must offer continual modification of SoS design and execution to compensate for accelerating complexity.
Holistic Dominance	The landscape of complex systems is dominated by the dynamically shifting impacts of technology, human, social, organizational, managerial, political, and policy dimensions. While it would be easier to deal with singular aspects of complex systems (e.g., technology), the realities suggest that the holistic range of factors must be taken into account. In addition, these impacts can, and will be, subject to changes over time in terms of their importance, interrelationships, and influence.	For system development, there must be an appreciation and accounting for the holistic spectrum of influences on system performance. The entire range of emergent technology, human, social, organizational, managerial, political, and policy factors must be considered. These factors, and their interconnections, must be included in development to improve the design and execution of system functions.
Information Challenge	It is an understatement to suggest that complex SoS is beset with exploding data and information. This is not new. However, the traditionally held relationships of data, information, knowledge, and wisdom must be questioned for continuing relevance and applicability for complex SoS. The structuring and ordering of expanding data confound complex SoS. Beyond increasing volumes, data challenges for complex SoS also include veracity issues, misinformation, and accessibility issues.	SoS development must be mindful of the flows and interpretation of information within and external to the SoS. Design and execution of SoS functions rely on information to perform. Thus, SoS development must include a focus on the two aspects: (1) ensuring that information is trustworthy and (2) design for the right information is available at the right place and the right time to support consistency in decision, action, and interpretation. In light of emergence, information access and timing become critical.

(Continued)

TABLE 9.1 (*Continued*)
Forces Impacting Complex Systems and Implications for Emergence

Forces	Explanation	Implications for Emergence Development
Contextual Influences	All complex SoS have a unique context within which they are embedded. This context includes the circumstances, factors, conditions, or patterns unique to the system. Context both enables and constrains a system. Similarly, the system constrains and enables the context. Separation of an SoS from its context is a false separation and only serves as a convenience for purposes of analysis. It is noteworthy that context is dynamic and will change over time as system knowledge evolves and the context experiences shifts.	SoS development that does not account for the context within which the SoS is embedded is deficient. SoS development must consider contextual influences on the SoS of interest, development execution, and expectations. Contextual considerations are not a one-time effort. Instead, context must be continually monitored, assessed, and accounted for during system development. Emergence can, and will, occur in context as it will in SoS. Without accounting for context, SoS development is incomplete.

The most basic articulation of emergence in SoS is the coming forth of patterns (behavioral, structural, performance) in real time as the SoS operates. These patterns cannot be known or predicted in advance. What is known with certainty is that patterns will emerge, although the precise form, timing, and impacts cannot be known in advance of their arrival. Thus, for complex SoS, the exploration of emergence is not directed to the constituent systems, but rather to the interaction and interrelationships among those systems. Emergence recognizes that unpredictable properties, behaviors, structure, and ultimately performance in SoS arise as the constituent systems operate within their local context. It would be the path of least resistance to 'do nothing' about emergence since it is going to happen and we do not have advanced knowledge as to when, where, manifestation form, or impact. However, this is not a palatable approach. Thus, our challenge is to pursue a deeper understanding and appreciation of emergence as a central phenomenon of SoS design, execution, and evolution.

The state of complex systems is consistent with Ackoff's (1994) depiction as 'messes'. Messes imply that we lack sufficient understanding to fully grasp the system, its context, or the approach to effectively move forward. Instead, we are left with complex situations that are appropriately referred to as 'wicked problems' (Rittel and Webber, 1973). Unfortunately, from these early articulations of difficulties facing complex systems, we appear to still be confounded in effectively dealing with complexity. It is noteworthy that emergence certainly plays a role in our difficulties, where unpredictable behavior/performance continues to dominate the complex system landscape. Traditional forms of analysis, stable planning, and deep understanding of a system are past relics that no longer seem capable of addressing complex systems and their constituent problems.

There is no better example of the difficult state of complex systems than the experience of catastrophic events. The examination of a catastrophe offers an ideal opportunity for a deeper exploration of emergence. The nature of a catastrophe represents the epitome of emergence. Catastrophic events are rare, harmful or disastrous (extensive loss of life or property damage), natural or manmade, and unexpected. The understanding of catastrophes occurs after experiencing the events and their aftermath. Thus, looking at a catastrophe can provide an opportunity to examine emergence in action and draw insights into associated phenomena, with the ultimate objective to better understand emergence and provide implications for practice.

In this chapter, we seek to provide a detailed exploration of emergence within the context of a catastrophe (Deepwater Horizon Disaster (DHD)). There are five major elements of this exploration (Figure 9.1). First, we examine the nature of emergence in SoS. A systemic perspective of emergence is established. This perspective draws on the roots of emergence in systems theory and is projected to the SoS domain and catastrophe. Second, we establish a systems-based framework for the examination of emergence in complex SoS. This framework is developed to provide a more rigorous accounting of the emergence phenomenon. The systemic framework for emergence is established as a means of classification, examination, and accounting for the multifaceted attributes of emergence for an SoS. Third, background and context for the DHD are developed. The nature of the unfolding disaster is captured, and the pivotal elements that occurred as the disaster evolved are examined. Fourth, the evolution of the DHD is examined using the systems-based emergence assessment framework. The particular conditions, events, and progression of the DHD disaster are viewed with respect to emergence. Emphasis is placed on the examination of the disaster in light of the emergence. A deeper understanding of the disaster and the implications

Implications: Development of the implications and guidance for emergence based on examination

Emergence in SoS: Examination of the relationship between emergence and SoS

Exploration: Examination of Deepwater Horizon from the emergence framework

Emergence Framework: Establishing a systems-based framework for the articulation of emergence

Deepwater Horizon: The background and context for the Deepwater Horizon disaster

FIGURE 9.1 Organization of the chapter.

of emergence occurring in the disaster are examined. Fifth, the set of conclusions and implications for practitioners who must confront emergence in complex systems on a daily basis are developed. Although the extraordinary circumstances of the DHD are an aberration from day-to-day operations of complex systems, they provide an opportunity to suggest an important set of practitioner guidance for more effectively dealing with emergence. The chapter closes with a set of insights for practitioners interested in a deeper sophistication in dealing with emergence in SoS.

9.2 EMERGENCE IN SYSTEM OF SYSTEMS

In this section, we explore the nature of SoS, including the classification, distinctions, and relationship of emergence in the context of SoS. Additionally, the nature of emergence is explored, setting the foundations for a deeper application of emergence in SoS through a systems-based framework.

9.2.1 THE SYSTEM OF SYSTEMS (SoS) CONCEPT AND EMERGENCE RELATIONSHIP

While there are different formulations of SoS and its counterpart System of Systems Engineering (SoSE), there is considerable agreement as to the essential characteristics for 'something' to be considered an SoS. These characteristics generally follow from the original work of Maier (1999) and Sage and Cuppan (2001), having been amplified in subsequent works (e.g. Gorod et al. 2008; Gorod et al., 2014). These characteristics are presented in Table 9.2.

In meeting these attributes, there are at least four types of SoS identified (Guide US DoD, 2008; Maier, 1999; Dahmann & Baldwin, 2008). With respect to the different forms, emergence is present in each of the forms and can be seen as potentially impacting the design, execution, and development of the different SoS forms.

1. **Virtual**: Lacking centralized management and integrated purpose and whose maintenance is largely unstructured with large-scale emergent behavior,
2. **Collaborative**: Voluntary interaction of constituent systems with a common purpose and establishment of expectations without direct power to enforce standards residing in the SoS,
3. **Acknowledged**: Common recognized SoS objectives and designation of integration management responsibility and resource allocation, although constituent systems retain a level of independence with changes being achieved through joint collaboration, and
4. **Directed**: A purpose built SoS intended to fulfill a specific intent, with central management where there is a subordination to the higher level SoS.

Irrespective of SoS type, emergence is a characteristic that is present. Additionally, as the SoS operates, emergence continually produces structural, behavioral, or performance shifts over time. This emergence may be 'intended' or 'unintended', but it is present, evolves, and must be dealt with as the SoS is established, develops, and is eventually disbanded.

TABLE 9.2

SoS Characteristics

SoS Characteristic	Description
Operational independence of constituent systems	Each constituent system has the ability and is explicitly designed to operate independently, supporting a mission unique to the system. Thus, SoS constituent systems have not been developed with explicit consideration for integration into an SoS for which they may become a part.
Managerial independence of constituent systems	The constituent systems comprising an SoS can be separately acquired and are independently managed. SoS constituent systems are largely autonomous with respect to decisions, actions, and interpretations that they engage. Incorporation into an SoS requires 'surrendering' of a degree of that autonomy to become part of an SoS that produces capabilities beyond that of any of the constituent systems.
Evolutionary development	SoS evolves over time. Constituent system capabilities may be added, removed, or modified as needs change and experience is gained. Evolution invokes a longer time horizon that suggests that SoS development is not revolutionary or instantaneous. Thus, SoS must take the 'long view' for development.
Emergent behavior	SoS have emergent capabilities and properties that do not reside in the constituent systems. Additionally, SoS emergent behavior spans the entire spectrum of constituent systems, their local environments, the interrelations with other SoS constituents, and the macro environment of the SoS. Sources of emergence may be internal to constituent systems/SoS or emanate from the local or macro-environment of constituents/SoS.
Geographical distribution of systems	SoS are comprised of constituent complex systems that are geographically distributed with the ability to readily exchange information. The recognition of geographical separation of SoS constituents recognizes that geographical separation and the associated information flows increase the complexity in an SoS.
Self-organization	SoS produces structural and behavioral patterns that did not exist and were not conceived by design. These patterns only emerge and are recognized as an SoS operates within its unique environment. Self-organization can be a purposeful approach to allowing SoS structure to evolve unconstrained, producing emergent behavior, structure, or performance. In contrast, SoS development can proceed with constraints that can produce 'intended' or 'unintended' emergence.
Adaptation	SoS continually evolves, either by proactive redesign or self-organizing processes, to maintain performance levels in response to external environmental shifts or internal shifts. Adaptation in an SoS provides for adjustment to shifts in the environment, context, or SoS itself. Ultimately, adaptation is about maintaining a sufficient 'fit' of the SoS such that it can continue to be viable in pursuit of its mission/purpose.

There are several important implications for emergence in SoS. First, SoS, and engineering of an SoS, is focused on the integration of multiple diverse constituent systems into a higher-level system. This higher-level SoS is capable of pursuing a purpose (mission) that exists beyond the capabilities of any of the constituent systems to achieve in isolation. Although there are 'intended' capabilities sought in an SoS design, emergence suggests the possibility that unpredictable consequences will appear, to be known only after the SoS is in operation. This does not suggest that the 'engineering' of the SoS was deficient. Instead, there are 'emergent' conditions, circumstances, events, or patterns that will occur despite our best efforts to understand the SoS. Second, in contrast to single systems, SoS escalates the complexity, and corresponding opportunities for emergence, exponentially. The formulation of previously autonomous constituent systems into an integrated SoS increases the complexity of the situation exponentially. With escalated emergence, more traditional approaches that have been geared to 'less' complexity (e.g., traditional systems engineering, Flood and Carson, 1993) are rendered suspect when applied to SoS. Increasingly complex circumstances of SoS suggest a high degree of emergence. Thus, emergence, which is characteristic of SoS, lies beyond the capabilities of traditional approaches to deal with complex SoS, and their associated emergence, through traditional approaches (e.g., Systems Engineering). The distinction between traditional approaches to 'systems' versus the demands of SoS amplifies the role of emergence for increasingly complex SoS.

Third, more traditional 'systems' problems have been effectively addressed (Keating, 2014; Katina, 2016) as well bounded, objectively formulated, and amenable to rigorous systematic resolution. With respect to emergence, these problems are less inclined to experience the consequences of emergence characteristic of more complex SoS. In contrast, SoS problems are ambiguously bounded, subjectively formulated, and more compatible with systemic-based approaches. SoS problems are more compatible with the perspective of emergence, where complete knowledge is not possible, is fallible, and will continue to evolve as an SoS operates within its environment. Fourth, there is not 'one size fits all' with respect to the formulation of SoS. Instead, there are an infinite variety and variability of SoS. Everything from an energy grid to the Internet of Things can be examined from an SoS perspective. This also implies that the continuing and growing presence of emergence in SoS is something that must be dealt with if these systems are to address needs or problems to advance societal aims. Fifth, there has been a propensity of traditional systems approaches (e.g., Systems Engineering) to focus heavily on the 'technical' aspects of complex system problems. While there are emergent conditions that can and will develop from the technical configurations of a system, it is naïve to falsely bound out other important aspects of a system/SoS from which emergence may be generated. For SoS, this implies that sources of emergence should be considered across the wide-ranging spectrum of not only technology, but also organizational, managerial, human, social, policy, and political dimensions. Thus, treatment of emergence from a SoS perspective calls for a much more 'holistic' set of considerations. The result of overemphasis on 'technology first, technology only' considerations for emergence in SoS is shortsighted. In many cases, emergence in technology/technical aspects of an SoS may be placed in the background, giving priority to 'non-technical' drivers of emergence wielding much more impact, positive or negative. In addition, even 'technology-based emergence' may cascade or be linked to other emergent events in an SoS.

Appreciation and incorporation of emergence considerations are central to SoS. SoS has been described as, 'The design, deployment, operation, and transformation of higher level metasystems that must function as an integrated complex system to produce desirable results' (Keating et al., 2003, p. 23). In a more recent evolution of SoS engineering, the notion of SoS being integrated by a 'metasystem' has matured to a current state of Complex System Governance (Keating and Bradley, 2015; Keating and Katina, 2016, 2019). The tenets of CSG are grounded in Systems Theory (laws governing the behavior and performance of complex systems) focused on integration and coordination of multiple systems, Management Cybernetics (the science of effective system organization) focused on provision of communication and control, and System Governance (essence of system direction, oversight, and accountability) focused on higher-level integrative design and evolution of systems. As emergence for SoS is considered, following earlier works of Keating and Katina (2018), the four primary products produced by an SoS, with emergence implications, include (and summarized in Figure 9.2):

- **Integration**: Integration permits constituent systems in an SoS to function as a 'unity'. This unity must provide the purpose/mission achievement that lies beyond any of the constituent systems to provide. As the structure,

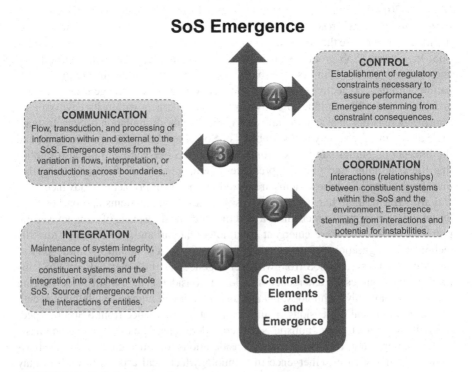

FIGURE 9.2 SoS elements and emergence.

behavior, and patterns emerge for an SoS, the ability to 'predict' the specific nature, form, and impacts of that emergence is questionable. It is remiss to believe that an SoS can be initiated without experiencing emergence.

- **Coordination**: Coordination is concerned with the interactions (relationships) between constituent systems within the SoS, and between the SoS and external entities, such that unnecessary instabilities are avoided. With increasing interactions comes the corresponding potential to experience emergence. For SoS, the elaboration of interactions (coordination) extends beyond the limited grasp of strictly technology-based exchanges that can be somewhat predictable in their presence and impacts. Instead, emergence stemming from interactions in support of coordination is unlikely to be either predictable in their nature or resulting impacts on the SoS. The escalating coordination requirements for SoS enhances the prospects for experiencing emergence. This is due to the increasing number and richness of interactions in an SoS.

- **Communication**: Communication involves the flow, transduction, and processing of information within and external to the system. As communication is achieved, the prospect for experiencing emergence increases. Communication in SoS is not restricted to the more traditional and narrow emphasis on technical communication (e.g., emphasis on interoperability). In contrast, communication is achieved through multiple channels, via multiple mechanisms, with a wide range of intents, and across a host of actors. Thus, there are many conditions for emergence in communications flow and interpretation of information in an SoS.

- **Control**: The primary function of control in any system is to invoke constraints (regulatory capacity) such that the system purpose/goal can be achieved. For SoS, the challenge for control is the 'preservation of the greatest possible level of autonomy for constituent systems'. In this sense, autonomy is the degree of freedom and independence of decision, action, and interpretation of actors in constituent systems. With respect to emergence, invoking constraint is a source of emergence, with potentially positive or negative impacts accruing.

Examination of an SoS presents an opportunity to explore the nature of emergence for this class of complex systems. SoS experiences emergence on a routine basis. The degree of complexity in an SoS suggests the inevitability of emergence. In addition, the exploration of a disaster in an SoS presents an opportunity to provide a 'deep dive' into understanding the precursor conditions, precipitative events/activities, the evolution of the emergence, and the post-emergent activities/actions in response. In the most general sense, a disaster is an event that has a sudden onset (e.g., accident or natural/manmade catastrophe) and has devastating consequences such as loss of life or severe damage to property. For examination of emergence in an SoS, a disaster provides a unique window into an event. Since serious disasters are well documented, the precipitating conditions, the actual event, the evolution of the event over time, and the response are well documented. Thus, exploration of emergence in a disaster becomes 'ripe' for analysis and detailed examination. Through the deconstruction of a disaster, the 'fog of emergence' can be lifted to provide analytical insights into the nature of emergence.

The inevitable challenge of retrospective analysis of emergence is that the richness of the ongoing situation can never be fully appreciated or examined. However, rich documentation and the accounting of events over the time horizon of the disaster provide as close a look at emergence as can be gleaned after the event. The sacrifice of immediate 'on the ground' observation of emergence is traded off for the ability to provide a detailed post-disaster examination.

Emergence is certainly a recognized characteristic of SoS and has been established as such in the literature (Bianchi, et al., 2015; Karcanias and Hessami, 2011). SoS has been described in a variety of ways (Keating et al., 2003, 2014; Keating, 2005; Keating and Katina, 2011), and attributes invariably include emergence as a central characteristic of SoS. Prior to the examination of emergence in SoS, it is important to provide a grounding background for emergence in general, and then to shift focus to target emergence specifically for SoS.

9.2.2 Emergence Essentials for SoS

Emergence is not a new concept for complex systems, or SoS, and has been a dominant theme in the literature for centuries. Early dating of the concept can be traced to Aristotle, who proclaimed that the whole is more than the sum of the parts (referenced Aristotle, 2002). Emergence has been part of the classical systems theory for some time. The basic suggestion of emergence from systems theory suggests that system properties (patterns, capabilities, performance, structure, behaviors) emanate from interactions between system elements and between system elements and the environment (Hitchins, 2008; Sousa-Poza et al., 2008). For complex systems, properties that emerge through interactions in a system are not traceable to any of the specific properties of constituent systems/elements and cannot be understood/predicted from the properties of those constituent systems/elements. This driving concept, although elaborated over time, has existed for centuries, having the grounding dating to Aristotle's proclamation that the whole is more than the sum of its parts (Checkland, 1999).

Emergence has evolved from very early descriptions. In fact, Holman (2010) traces the roots of the concept of emergence to original work by Lewes (1875, p. 412) who suggested that '[T]here is a co-operation of things of unlike kinds. The emergent is unlike its components . . . and it cannot be reduced to their sum or their difference'. More recently, in the inaugural issue of the journal *Emergence*, the concept was captured as

> ...the arising of novel and coherent structures, patterns, and properties during the process of self-organization in complex systems. Emergent phenomena are conceptualized as occurring on the macro level, in contrast to the micro-level components and processes out of which they arise.

> *Goldstein (1999, p. 49)*

This perspective was important in recognizing that emergence has a relationship to the system principle of self-organization, a principle directed to the process that generates 'emergent' structure, behavior, patterns, or performance in complex systems that come about as a system operates. The notion of emergence as providing novelty is suggested by Lucas and Milov's (1997) observation that emergence requires the development of new categories. These categories are not present in existing elements (parts) and require a new vocabulary to describe properties that did not previously exist at the part/element level of complex systems. Chalmers (1990) suggests that emergence in design occurs when a system is designed in accordance with certain principles and expectations, yet 'interesting' properties arise that were not intended or conceived in the original design. This amplification of the unintentional nature of emergence is important, also from the perspective of being both positive or negative in impacts.

Irrespective of the prevalence and development of emergence in the literature, there are many different perspectives without a dominant 'definition' or consensus concerning the phenomena associated with emergence. However, the development of several central themes concerning emergence is instructive for our present investigation. Table 9.3 is a set of emergence characteristics drawn from a broad cross-section of literature (Adams et al., 2014; Kim, 1997; Keating, 2009; Holman, 2010; Goldstein, 1999; El-Hani and Pereira, 1999; Corning, 2002; Arshinov and Fuchs, 2003; De Wolf and Holvoet, 2004). While not presented as the definitive set, or absolutely complete, the set provides a solid grounding of commonly held aspects for the concept of 'emergence' in the literature.

TABLE 9.3
Common Attributes of Emergence

Emergence Characteristic	Description
Novelty	Properties or features that did not previously exist. The nature of these properties is the source for the unpredictability of emergence.
Coherence or Correlation	Emergent characteristics maintain a sense of identity over time. A higher-level unity is maintained from correlation of lower-level elements.
Global level	Properties exist at the higher (global/macro) level as opposed to the lower level (micro) from which the properties emerge.
Dynamic	Properties evolve over time and are not capable of being predetermined or predicted in advance of their manifestation.
Ostensive	Recognition of emergent properties becoming known when they become exposed – never the same as previously exposed emergent properties/phenomena.
Interacting parts	Production of emergent properties requires that elements interact.
Decentralized control	There is no central control. Macro-level behavior is produced by local mechanisms that interact to produce global-level behaviors.
Downward causation	Micro to macro level produces emergent behavior/structure. Macro to micro emergent structure influences the micro level generating causation in a downward direction.
Robustness	Single entities cannot be a single point of failure, as emergents are insensitive to perturbations. This generates robustness.
Flexibility	Degradation due to perturbation/error is gradual, meaning that flexibility permits the emergent structure to remain.

There are multiple formulations of emergence (Fromm, 2005; Plowman, et al., 2007: Kalantari et al., 2020). In the examination of emergence, Holland (1998) suggested it is unlikely that a topic as complicated as emergence will submit to being captured in a single definition. However, for the current exploration, we point out two important emergence classifications. First, the classification of 'strength of emergence'. Stemming from the work of Chalmers (2006), *strong emergence* occurs when higher-level phenomena emerge from a lower-level domain but are not deducible from that domain. This focuses on novelty experienced as irreducible (incapability of understanding by reduction to phenomena held at a lower level) and downward causation (novel properties which arise at the macro level and generate causation impacting behavior at the micro level). Thus, novel properties arise in a higher-level system which are not attributable to lower-level constituent systems but yet can impact the behavior of the constituents. Chalmers (2006) goes on to describe *weak emergence* as existing when a novel property unpredictably/unexpectedly arises in a higher-level system. However, for weak emergence, although the phenomenon might exhibit unpredictable results, it may nevertheless be deducible given sufficient knowledge (initial state of the system and governing rules). Thus, additional knowledge may in fact shed light on the emergence. Additionally, emergence can be described as *intended* or *unintended* emergence. Sauser et al. (2009) suggest that for intended emergence the broad precepts of desirable (emergent) results from modifications in the system. However, unintended emergence is the experiencing of properties that are neither foreseen nor desirable within a range of anticipated outcomes. This is commonly referred to as 'unintended consequences'. Quite simply, intended emergence occurs when systems are perturbed with specific and desirable 'emergent' outcomes in mind. Although the 'precise' nature of the emergence is not fully known, the range of expected system response is known and anticipated. This lends itself to more rigorous analysis, modeling, and exploration of expected system reactions to perturbations. In contrast, *unintended* emergence recognizes that the emergent properties are not anticipated and could not be predicted in advance. Notwithstanding this strong-weak emergence, and intended-unintended distinctions, for our purposes, we focus on three critical aspects of emergence: (1) emergent properties not being held at the level of constituents; (2) the nature of the emergent properties being novel and not capable of prior 'knowing' their nature, form, location, or timing, or impacts; and (3) the inability to 'reduce' understanding of emergent properties to constituent entities from the level where they were produced. Thus, with respect to SoS, we offer a common perspective of emergence as a novel higher-level SoS property arising from interacting lower-level constituent systems, being incapable of precise prediction or reduction to properties of lower-level constituent systems.

In this section, we have: (1) identified the characteristics that delineate an SoS, (2) examined the nature of SoS and the role that emergence plays in SoS, (3) examined the central tenets of SoS and the role emergence plays in those tenets, and (4) suggested the opportunities presented to examine emergence in the context of SoS disasters. Given these perspectives of emergence, we now shift our development to generate a framework for emergence in SoS. This framework is grounded in systems theory and developed for application to better understand the nature of emergence in

SoS and facilitate application to operational settings. Ultimately, the intention is to provide a foundation that gives practitioners a better frame of reference and approach to deal with increasingly complex systems and their inevitable emergence.

9.3 EMERGENCE FRAMEWORK FOR SYSTEM OF SYSTEMS

In this section, we focus on the development of a framework for emergence in SoS. The development is approached in three stages. First, the foundations of a systems perspective of emergence in complex systems are examined. Second, a systems-based framework for emergence in complex SoS is articulated. Third, the implications for deployment of this framework for application in practice are suggested, with emphasis on the examination of emergence in SoS disasters.

9.3.1 SYSTEMS PERSPECTIVE FOUNDATIONS FOR EMERGENCE IN COMPLEX SYSTEMS

Emergence has been well recognized as a characteristic of complex systems (Keating and Katina, 2018, 2020) and a critical attribute of SoS. In the explanation of emergence, it has been described in various ways. Fromm (2005, p. 4) has described emergence as 'an effect or event where the cause is not immediately visible or apparent'. There are two important amplifications of this definition. First, description of emergence as an effect or event. This is important in that it suggests dual roles of emergence. An event is a specific occurrence that can be identified as emergent. However, as Holland (1998) suggested an effect is the impact, or consequences, of an event. Thus, these two aspects of emergence, event and effect, are important in developing a perspective to ground emergence.

A cogent description of emergence is provided by The Lloyds register as, 'emergence as a property of a system, can be used to describe relevant attributes that allow systems to withstand, respond, and or adapt to vast range of destructive events' (Lloyd, 2015, p. 7). With respect to robustness, this is descriptive of the range of 'emergent events' a system design can absorb. If the event(s) are beyond the system design capacity to absorb, the system will in the best case experience performance degradation, and in the worst-case experience outright failure. The related concept of resilience is captured as the ability of a system to return to the original, or near original, state of performance following a perturbation. Consistently, resilience has been defined as 'a process linking a set of adaptive capacities to a positive trajectory of functioning and adaptation after a disturbance' (Norris et al. 2008). Thus, emergence capabilities can be linked to capabilities that enable complex systems (i.e., individuals, communities, institutions, businesses) 'to survive, adapt, and grow no matter what kinds of chronic stresses and acute shocks they experience' (Martin-Breen and Anderies, 2011). These shocks can be considered as emergence. Emergence's relation to self-organization can be explained in the suggestion that complex systems organize themselves (Ashby, 1962; Skyttner, 2005). In this case, 'organizing themselves' is an inherent capability that somehow 'emerges' to influence structural and behavioral patterns in a complex system, primarily resulting from the interaction among parts of a complex system.

In taking a systems perspective of emergence, we can suggest several themes that can be instructive in developing a perception and subsequent framework for emergence. These themes from a systems worldview on emergence include:

- **System Impacts Stemming from Emergence That Produces Unabsorbed Variety**: Requisite variety was developed by Ashby (1957; 1991) to explain that a system must have sufficient regulatory capacity to match or exceed the variety being generated by the environment. Variety being generated from the environment can be considered 'emergent events' with potential impacts on a system. Thus, regulatory capacity is a function of system capability to mount an effective response(s) to disturbances (emergence) such that essential variables necessary for sustained system performance are maintained within desirable limits. If variety stemming from emergence is left unabsorbed, then system performance is degraded.

- **Robustness of a System to Compensate for 'Classes' of Emergent Events and Their Effects**: Robustness is a function of the design of a system. Robustness indicates the range of disturbances (emergent events) over which a system can mount an effective response without redesign of the system. Thus, although the precise nature, form, and timing of an emergent event cannot be known in advance, a system (design) has a range of pliability in effectively dealing with emergence. The degree to which a system can compensate for emergent events is an indicator of system robustness. Robustness can be enhanced through the active redesign of a system, such that future episodes of 'similar' emergent events can be compensated for through the modified system design.

- **Resilience of a System to Return to Prior Configuration and Performance Levels Following Emergent Events**: At the most foundational level, resilience has been articulated as 'the act of rebounding or springing back' (Murray et al., 1977). For complex systems, subsequent to experiencing an emergent event, resilience suggests the ability of a system to respond to externally generated perturbation(s) [emergence] such that performance is maintained. The degree to which a complex system can 'bounce back' from emergent events is a function of the resilience of that system. Resilience to emergent events is related to the robustness in a system design, taking on the added element of 'execution' of that design to compensate for emergent events.

- **Viability of a System as a Function of Ability to Continue Existence in the Face of Emergence**: Viability is a concept that stems from the field of management cybernetics (Beer 1979, 1985). In essence, viability is the ability of a complex system to continue existence. With respect to emergence, viability is an indicator of the debilitating impact of an emergent event. If the system is sufficiently robust and resilient in design and execution, then viability will be maintained. In contrast, viability (existence) will be compromised if the system cannot mount an effective response to an emergent event(s). An important note is that viability is not a binary proposition of a system existing or not existing. The degree of impact for an emergent event will impact viability, and although the impact may not result in catastrophic failure, it can diminish performance to varying degrees.

- **Antifragility of a System to Evolve in Response to Emergent Events**: Antifragility follows the work of Taleb (2012). This concept suggests that systems can thrive under conditions of disturbance, and thus be antifragile to external/internal perturbations (emergence). From a systems perspective, the generation of antifragility comes from response to perturbations (emergence) in ways that strengthen the system execution (first-order change), system design (second-order change), or a combination of both. To the degree that a system can evolve to become increasingly antifragile in the face of emergence, it will move toward continuing levels of performance as it meets challenges of changing environments.
- **Control as Providing Regulatory Constraint That Enables Sufficient Autonomy Necessary to Confront Emergence**: Control (Keating and Katina, 2019) is focused on the establishment of constraints that provide regulation necessary to ensure consistent performance and future system trajectory. Control is essential to ensure that the system stays on a trajectory that will provide future viability in response to changing conditions and circumstances. In effect, system controls provide capability to continue performance in the face of emergent events and their impacts. Limiting controls to only those that are essential preserves autonomy (freedom and independence of decision, action, and interpretation) through 'minimal' regulatory capacity. Thus, system elements are in a better position to respond appropriately to emergent events.
- **System Identity as a Reference Point and Guide for Dealing with Emergence**: System identity provides the 'essence' of what constitutes a system (Beer, 1979, 1985; Keating and Katina, 2019). It provides for continuity in the face of turbulence experienced as emergent events. The degree to which a system has a strong identity is a source of continued viability (existence) as increasing emergent events challenge a system. The maintenance of a strong identity is one factor that helps ensure system performance can be maintained in the face of emergent events. A strong identity provides for consistency in decision, action, and interpretation in response to challenges incurred from emergent events.

Based on a systems perspective of emergence, we add an additional dimension to the strong-weak and intended-unintended descriptors. This systems-based depiction of emergence is captured as *absorbed-unabsorbed* emergence. Absorbed emergence is accounted for by the existing design and corresponding execution of a system. Absorbed emergence does not have a negative or positive impact on system performance. Instead, absorbed emergence simply notes that the system continues to function, producing expected behavior/performance, in the face of emergent events. In this sense, absorbed emergence is addressed by the design and execution of a system, without experiencing negative consequences. In contrast, unabsorbed emergence suggests that system behavior/performance is impacted negatively stemming from emergent events. The design and execution of a system are not capable of mounting

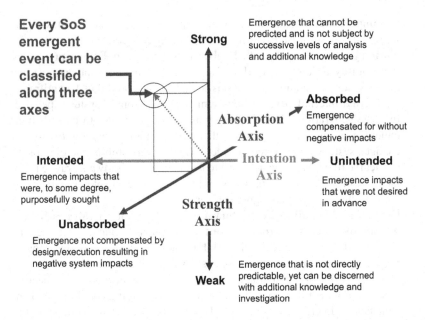

FIGURE 9.3   Emergence classification of events.

a response that can thwart the impacts of an emergent event(s). The result is degradation of system performance in the best-case scenario. In the worst case, the result might entail system collapse/failure. The absorbed-unabsorbed classification can be added to the previously designated weak-strong and intended-unintended classifications of emergent events (Figure 9.3). The classification of emergent events helps to understand the implications for emergence planning, emergence response development, and system design/execution development in response to emergence.

9.3.2   A SYSTEMS-BASED FRAMEWORK FOR EMERGENCE IN COMPLEX SOS

Based on the development of the systems perspective for emergence and the corresponding articulation of emergence, a systems-based framework can be developed. This framework serves to support several objectives with respect to the application and practice related to emergence. Among these objectives are included:

1. **Grounding of the Framework in Both Systems Theory and Emergence Literature**: The framework is based on the stability of a strong underlying conceptual/theoretical set of underpinnings. As such, the framework has both broad applicability and maintains a linkage to a strong and stable basis in emergence and systems theory.
2. **Explanatory and Predictive Enabling Capabilities to Better Understand the Genesis, Evolution, and Mastery of Emergent Events**: The framework operates across three time horizons for dealing with emergent events, including

pre-emergence (prior to experiencing the emergent event), intra-emergence (as the emergent event is unfolding), and post-emergence (after the emergent event has concluded). Thus, the framework can support predictive capabilities as well as explanatory guidance for emergent events.

3. **Practical Implications for Dealing More Effectively with Emergent Events**: The implications of the framework concerning emergent events cross the spectrum of design, execution, and development. Design can be assessed with respect to the degree of robustness and resilience for emergence. Execution can be examined for immediate responses in the face of unfolding emergent events and assuring continued viability of the existing system. Development can be assessed with respect to implications related to modifications to design or execution to craft a system that is more anti-fragile to the impacts of future emergent events.

4. **Practitioner-Based Guide to Applications Related to Emergent Events**: For practitioners facing current- and future-focused emergence in complex systems, the framework provides a guide. This guide offers a rigorous treatment of emergence and directions for clarity of understanding and potential paths forward to deal with emergent events.

9.3.3 A Systems-Based Framework for Emergence

Given the background and perspectives developed for emergence in complex SoS, we have developed a framework. This framework is intended to provide a formulation of: (1) emergence in complex systems and SoS, (2) analysis for classification of emergent events, and (3) cues with respect to more effectively dealing with emergence. Figure 9.4 provides a high-level view of the framework. Each stage of the framework will be expanded in the subsequent text.

The Systems-Based Framework for Emergence is designed to guide the assessment of emergent events occurring in a system under specific conditions at different stages of evolution of the event. For conciseness, Table 9.4 provides a detailed overview of the six major stages of the framework. There are several assumptions upon which the framework is based, including: (1) the framework is intended to be systemic (applied from a systems perspective), (2) the framework is iterative in nature such that it is cyclically and iteratively applied and does not follow a strict linear approach, (3) the framework supports emergent event classification as a precursor to the development of a response, (4) the framework suggests response development being informed from the primary stages (System, Conditions, Event, and Stage), and (5) development of feasible responses in light of the emergent event Classification and Stage of the emergent event.

This emergence framework is not presented as the complete or only approach to examine emergence in complex systems. To do so would be overly naïve, ambitious, and considerably overstated. However, the framework does provide a comprehensive treatment of emergence as well as a set of considerations that can guide more effectively dealing with emergent events over their life cycle. It moves beyond superficial accounting of emergence in complex systems.

Emergence Framework

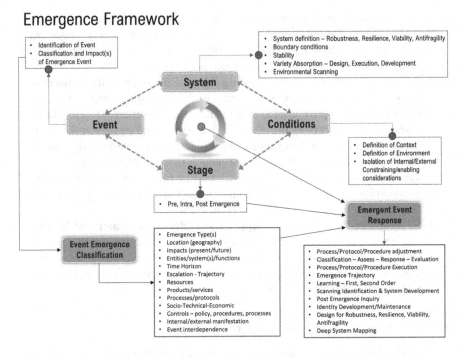

FIGURE 9.4 Emergence framework for complex systems.

Having established the emergence framework, we shift attention to implications for practice. The DHD serves as a platform to examine the implications of an emergent event in light of the emergence framework.

9.4 DEEPWATER HORIZON DISASTER

The DHD serves as an excellent case for examination of the Emergence Framework for application. The selection of DHD was made for several reasons. First, the DHD involved a scenario that was initially identified as a technical failure but in actuality crossed the spectrum of other aspects contributing to, and impacted by, the technology failure. Second, the DHD is well documented with respect to the precipitating conditions, escalation trajectory, and post-disaster implications. Third, the complexity of the DHD is such that simple cause-effect relationships are insufficient to explain the disaster. In this section, we provide a short overview of the DHD. The details are only covered to the degree necessary to convey the information necessary to support the examination and implications for emergence in the situation.

The DHD stands as the worst oil-spill incident in the history of the petroleum industry. Irrespective of the economic damage, DHD resulted in 11 fatalities and 16 serious injuries. Additionally, 134 million gallons of oil spilled into the Gulf, creating an environmental and economic catastrophe. From the commission on DHD (Graham et al., 2011, p. xiii):

TABLE 9.4
Description of the Emergence Framework Stages

Framework Stage	Description
System	The definition of the system includes a mapping of the system across input, transformation, outputs, feedback, boundary conditions, and environment. Establishment of the stability of the system (degree to which the system is in balance, and capable of absorbing variety generated from the environment) in relationship to design and execution in relationship to the environment perturbations (emergence). Emphasis on the capability of the system to perform environmental scanning to identify existing and potential sources of emergent events.
Conditions	Conducting a detailed examination of the system context (factors, circumstances, conditions, patterns, or trends that enable or constrain a system). Examination of the environment within which the system is embedded, including inputs to the system (e.g., resources, demands) and outputs to the environment (products, services, or information) produced and consumed as value in the environment. The identification of priority considerations in the context or environment that can enable/constrain system performance.
Stage	Definition of the specific stage for the emergent event. The stage can be pre-emergent (identified prior to the occurrence of an event), intra-emergent (as the event and its immediate aftermath are unfolding), or post-emergent (after the event and ramifications stemming from the event are complete). Each stage has different implications for analysis and treatment. In effect, the three stages define the life cycle of an emergent event from preconception through completion.
Event	The actual emergent event(s). This can also include ancillary or cascading events that may be interrelated to the precipitating event. Emergent events should be identified in sufficient detail to permit their classification and preparation for further analysis.
Classification	The emergent event must be classified along a set of defining dimensions. Among these dimensions are: Emergence Type (positive-negative, intended-unintended, absorbed-unabsorbed), Location (the geographical location of the event), Impacts (the effects and implications of the event for system performance for present and future), Entities/System (the specific entities/systems that are engaged in the emergent event), Time Horizon (the time for occurrence and unfolding of the event), Escalation (the propagation of the event in scope, trajectory, and impact for the system), Resources (the impact of the event for system resources), Processes/protocols (the formal directives in place to guide the system with respect to the emergent event), Socio-technical-economic (the attributes that demark the event across social/human, technology/technical, and economic dimensions), Controls (the regulatory mechanisms that are in place and relevant to dealing with emergence events), and Event Interdependence (the degree to which the event can be considered in isolation from other events or related aspects). It is important to note that the classification of an emergent event may shift over time as new knowledge and understanding materialize.

(Continued)

TABLE 9.4 (*Continued*)
Description of the Emergence Framework Stages

Framework Stage	Description
Response	The specific actions that are taken in response to the emergent event. These responses may be taken at any stage of the emergent event life cycle. Among possible responses are Process/Protocol/Procedure adjustments (shifting the formal ways of dealing with events over their life cycle), Classification – Assessment – Response – Evaluation (the development and deployment of responses over the life cycle of the emergent event), Process/Protocol/ Procedure execution (the engagement of existing approaches to deal with the emergent event), Emergence Trajectory (purposeful management of the path of the emergent event), Learning (engagement of detection and correction of errors stemming from the event – crossing first-order changes to execution and second-order change for modifications to the system design), Scanning Identification and System Development (engaging improvement in the ability to detect and process emergent events), Post-Emergence Inquiry (conducting exploration of emergent events after their conclusion), Design for Robustness, Resilience, Viability, Antifragility (engaging system redesign to enhance the classes of emergent events capable of being accommodated, providing capabilities to return to prior states of performance following emergent events, maintenance of system existence through engineering of 'variety', and capability to enhance a system such that it comes back 'stronger' [antifragility] following an emergent event), Deep System Mapping (construction of a comprehensive mapping of a system to locate the precise occurrence, positioning, and impacts of an emergence event).

On April 20, 2010, the 126 workers on the BP Deepwater Horizon were going about the routines of completing an exploratory oil well—unaware of impending disaster. What unfolded would have unknown impacts shaped by the Gulf region's distinctive cultures, institutions, and geography—and by economic forces resulting from the unique coexistence of energy resources, bountiful fisheries and wildlife, and coastal tourism. The oil and gas industry, long lured by Gulf reserves and public incentives, progressively developed and deployed new technologies, at ever-larger scales, in pursuit of valuable energy supplies in increasingly deeper waters farther from the coastline. Regulators, however, failed to keep pace with the industrial expansion and new technology—often because of industry's resistance to more effective oversight. The result was a serious, and ultimately inexcusable, shortfall in supervision of offshore drilling that played out in the Macondo well blowout and the catastrophic oil spill that followed.

Table 9.5 provides a summary of the preliminary findings and observations related to the DHD. This table is based on the findings from the Deepwater Horizon Interim Report from the National Academy of Engineering and the National Research Council (National Academy of Engineering, 2010). This set of findings and observations is insightful in setting the stage for the exploration of emergence in the DHD. However, care must be taken with respect to the post-event analysis. In particular: (1) post-event analysis largely removes the 'fog of emergence', permitting deep analysis

TABLE 9.5

Findings and Observations Related to the DHD

Observation/Finding	Description
Decision to abandon temporary exploratory well	Irrespective of test results that did not support the decision to abandon the exploratory well, the abandonment decision was made. The well integrity was not sufficient to support the abandonment of the well. Despite these indicators, the temporary well was abandoned. This decision failed to permit an effective barrier to hydrocarbon flow.
Actions for well-control of the hydrocarbon flows were not identified in a timely manner resulting in delayed response	The decision delays in proceeding with temporary abandonment were compounded by two items: (1) the delays in recognition that hydrocarbons were flowing into both the well and the riser (the pipe that carries the subsea oil and gas from the wellheads on the seabed to the floating unit on the surface), and (2) the failures (technical) in the Blowout Preventer (unit to prevent an oil spill for well safety).
Insufficient consideration of risks and lack of operating discipline	Decisions made indicated lack of knowledge by personnel which reduced the available margins of safety. This failed to take into account the complexities of the design and changes in the well plan.
Poor decisions related to the operation and safety of the well	Several contributory decisions precipitating the event, including: (1) changing supervisory personnel just prior to critical operational phases, (2) technical issue in not accounting for differentials in fluid pressures in cement density, (3) material selection issues for the uncased section of the well, (4) failure to adhere to modeling results for technical guidance of construction, and (5) several technical and operational decisions concerning well safety.
Insufficient 'checks and balances' for decisions	The schedule for completion of well abandonment procedures as well as the well safety considerations was lacking appropriate oversight. The resulting decisions were implemented without sufficient 'checks and balances' to provide oversight necessary to test decision adequacy and avoid a single point of decision failure.
Material integrity of the well stemming from selection of the cement and curing time	The type and volume of cement used, as well as the appropriate curing time for well abandonment, were in question as to their impact on support for well abandonment.
Failure in the technical solutions to address the well blowout	The Blowout Preventer was initiated late and failed to control the spill. Also, other features of the blowout preventer did not permit proper abandonment of the well (e.g., emergency disconnect system).
Failure of alarms and safety systems to operate as intended	Failure in these safety features was suspected as reducing the time available for evacuation of personnel from the site.
Failure of the approach for managing risks, uncertainties, and dangers in deepwater drilling operations	The design and operation of the deepwater drilling operations were inadequate to manage risks and inherent dangers. This was indicated by prior 'near misses' from which the operations and execution failed to learn.
Lack of a systems approach to provide integration of multiple different factors for well safety	The multiplicity of different factors that must be considered to address well operations and safety considerations were not adequately taken into account. These factors were not sufficiently treated from a systems perspective.

without the full ranging context and richness of the event(s) as they unfolded, (2) retrospective analysis is always void of the transient local context of 'real-time' decisions that were made without the possibility of unconstrained resources or time permitted for decision support, and (3) after the event, logical and rational analysis is used to frame the situation – missing the 'messiness' inherent in the unfolding disaster where parsing of the situation creates the ability to isolate elements that are not possible during the evolution of the disaster. Nevertheless, with respect to facilitating a better understanding of emergence, and implications for enhanced practice, the exploration of the DHD offers a rich source for analysis.

With this background on the DHD, we examine emergence in relation to the evolution of the disaster emanating from the precipitating event (explosion of the well).

9.5   EMERGENCE IN THE DEEPWATER HORIZON DISASTER

The DHD provides an opportunity to examine the nature, role, and implications of emergence for an actual setting. The examination of emergence is amplified given the catastrophic nature of the disaster and the documented events. These events span the spectrum of emergence, which unfolded over a long time horizon and was well documented. In this section, we examine emergence in the DHD and focus on the exploration of several key points. First, the actual event is examined. This event is actually technical in nature. Second, the three stages of emergence surrounding the event are examined. These phases include *Pre-emergence* factors that occurred prior to the actual event and were considered to be contributory to the event. *Intra-emergence* is focused on the unfolding and evolution of the situation following the event. The focus of this stage is on the evolution of emergence as following an 'emergence trajectory' as the situation moves through to 'completion' of the event and direct aftermath. *Post-emergence* is focused on the retrospective analysis and modifications/ recommendations. This permits more deliberate analysis removed from the ongoing pressures associated with 'getting through the disaster'. Retrospective analysis permits a deeper examination of central issues related to emergence. However, lost in this analysis is the richness of the ongoing unfolding of the event in real time. Thus, dealing with emergence in real time lacks the luxury of post-event analysis. Post-event analysis can be made without the pressures and urgency demanded of decision-action sequences responding to emergent events in real time.

For this examination of emergence, we are following the major categories of the framework, including Classification of Emergence (examination of positive-negative, intended-unintended, and absorbed-unabsorbed), Stage (pre-emergence, intra-emergence, and post-emergence), and Response. For emergence in the DHD, Table 9.6 summarizes emergence in the situation. The content for this table is derived primarily from Graham et al. (2011) as supplemented with emergence interpretations.

From Table 9.6, there are three primary conclusions that can be made with respect to emergence in the situation. First, it is noted that the 'bulk' of emergence types fall into the Negative – Unintended – Unabsorbed category. This suggests that the emergence identified was primarily negative in nature. Thus, the impact of the emergence resulted in negative consequences experienced. Additionally, the emergence was classified as Unintended. The results of the emergence event and subsequent impacts were not intentional. This suggests that individuals making decisions that led to the emergence events had an 'emergence blind spot' – making trade-off decisions that were expedient and not necessarily well thought out with respect to the potential for emergent conditions that might introduce negative consequences. Irrespective of the 'good intentions', the resulting negative consequences were suffered as a result of the decisions made. Interestingly, these decisions were not all made in real time but were made with considerable leeway in time prior to the emergent event. However, they created conditions that contributed to the occurrence of the emergent event. Also, the bulk of emergence was 'unabsorbed' by the system design and execution. This permitted the emergence to manifest itself in an unconstrained fashion, producing impacts that were detrimental to the safe operation of the system and setting an untenable failure trajectory.

Emergence was spread across the different stages of emergence (pre-, intra- post-). It is noteworthy that the pre-emergent conditions were difficult in trying to make an immediate and direct linkage to the negative emergence impacts to which they contributed. The linkage had to be inferred and was not directly observable. In contrast, intra- and post-emergent stages were more readily accessible, and thus emergence impacts were able to more readily be directly attributed. However, the three stages of the 'emergence life cycle' appear beneficial in characterizing emergence events as to their particular stage of occurrence, manifestation, impact, and response. The intra-emergence stages were often attributed to the results of the pre-emergence stages, creating the conditions under which the emergence could be experienced. However, direct evidence of this contribution to emergent conditions is not discernable in any direct way. Nevertheless, the existence of the pre-emergence conditions is instructive in examining the creation of circumstances favorable to emergence.

The emergence stage that seemingly is the easiest to navigate is the post-emergence stage. This stage has the advantage of 'hindsight' and drawing insights from more deliberate analysis. There are two emergence pressures that are not present in the post-emergence stage that impact pre- and intra-emergence stages. First, in pre-emergence stages, there is not the advantage of 'seeing connections' and implications for invoking, or not invoking, certain decisions/activities. The allocation of scarce resources to enhance a system is not done with the more precise after event knowledge held at the post-emergence stage. In addition, there is the possibility of complacency with the status quo, not realizing the potential for catastrophic consequences for engaging, or not engaging, different actions related to a system. Also, in the intra-emergence stage, there are pressures for making decisions on a more intuitive level, without the benefit of additional research or deliberate deep examination. Thus, intra-emergence decisions are apt to be more 'seat of the pants' decision – action – interpretation sequences.

In response to the DHD, there were several recommendations that occurred in the post-emergence stage. Interestingly, while these recommendations appear to be sound in their reasoning and analysis, they do not provide the coverage at the 'ground level' of the system. Instead, they are left at a very high level. One must assume that the ground-level system modifications were left to be undertaken by the owners of the system. Table 9.7 provides a summary of findings and recommendations in response to the investigation of the DHD (Report to the President, Graham et al., 2011).

TABLE 9.6
Emergence in the DHD

Emergent Behavior/Properties/Events	Classification (Positive-Negative, Intended-Unintended, Absorbed-Unabsorbed)	Stage (Pre-, Intra-, Post- Emergent event)	Associated Manifestation	Generating Process/Mechanism
Macondo well-explosion	Negative	Intra	Blowout of the well	Lax safety culture; failures in risk management precipitating the blowout
Trade-off of risk for exploration and production	Negative – Intended – Unabsorbed	Pre	Realization of emergent risk event that the government and industry were not adequately prepared to muster a robust and resilient response	Lack of adequate risk management processes and procedures to deal with the evolving disaster
Regulatory oversight reforms initiated for leasing, energy exploration, and production; immediately following and post disaster	Positive – Intended – Absorbed	Post	Reforms put in place to preclude 'similar' deep sea oil drilling disasters	Recognition that structure of regulatory oversight, internal decision-making processes free of political interference, appropriate technical expertise and environmental protection concerns
Insufficiency of self-regulatory behavior by the oil and gas industry concerning the safety of operations	Negative – Unintended – Unabsorbed	Pre	Lack of internal 'self-policing' resulted in complacency and lack of rigorous controls to either prevent or adequately respond to disasters.	Post review of disaster identified insufficient capacity to effectively implement and manage 'self-policing' mechanisms

(Continued)

TABLE 9.6 (*Continued*)
Emergence in the DHD

Emergent Behavior/Properties/Events	Classification (Positive-Negative, Intended, Unintended, Absorbed-Unabsorbed)	Stage (Pre-, Intra-, Post- Emergent event)	Associated Manifestation	Generating Process/Mechanism
Limited technologies, laws and regulations, and practices related to preventing and containing associated risks with respect to oil spills stemming from offshore deepwater drilling	Negative – Unintended – Unabsorbed	Pre	The emergent disaster demonstrated flaws in oversight and preparedness to effectively deal with the initial occurrence and quickly evolving circumstances	Insufficiency in the regulatory controls, practices, or technologies to mount an effective response
Environmental damage from disaster	Negative – Unintended – Unabsorbed	Intra, Post	Sensitive environment in the Gulf waters and coastal habitats lacked sufficient understanding of the associated risk to these phenomena	Ecological damage from emergent events and ongoing inabilities to cope with technological inadequacies to keep pace with the evolving crisis
Undermining of public confidence in energy sector, government regulators, and national response capabilities	Negative – Unintended – Unabsorbed	Intra, Post	Ongoing crisis without confidence that the industry or government could effective management the crisis	Inability to manage and bring under control the disaster
Flow of hydrocarbons into the well and riser	Negative – Unintended – Unabsorbed	Intra	Delays in recognizing that hydrocarbons were flowing into the well and riser and by a failure to take timely and aggressive well-control actions	Decision to proceed to the temporary abandonment of the Macondo well

(Continued)

TABLE 9.6 (*Continued*)
Emergence in the DHD

Emergent Behavior/Properties/Events	Classification (Positive-Negative, Intended-Unintended, Absorbed-Unabsorbed)	Stage (Pre-, Intra-, Post- Emergent event)	Associated Manifestation	Generating Process/Mechanism
Blowout preventer ineffective once actuated	Negative – Unintended – Unabsorbed	Intra	Failure of the blowout preventer to effectively seal the oil well	Blowout preventer failed due to unrecognized pipe buckling
Missed indications of hazard in decision to abandon the Macondo well	Negative – Unintended – Unabsorbed	Intra	Numerous decisions to proceed toward abandonment despite indications of hazard, such as the results of repeated negative-pressure tests	Insufficient consideration of risk and a lack of operating discipline
Adequacy of operating knowledge and emerging decisions of key personnel	Negative – Unintended – Unabsorbed	Pre, Intra	Net effect of these decisions was to reduce the available margins of safety that take into account complexities of the hydrocarbon reservoirs and well geology discovered through drilling and the subsequent changes in the execution of the well plan	Decision making with respect to safety margins for the temporary abandonment of the oil well
Changing key supervisory personnel on the Deepwater Horizon Mobile Offshore Drilling Unit just prior to critical temporary abandonment procedures	Negative –Intended – Unabsorbed	Pre	Key personnel not operationally 'sound' in dealing with abandonment issues and potential safety hazards	Insufficient skills, knowledge, and abilities to provide oversight of complex abandonment of the oil well

(Continued)

TABLE 9.6 (*Continued*)
Emergence in the DHD

Emergent Behavior/Properties/Events	Classification (Positive-Negative, Intended-Unintended, Absorbed-Unabsorbed)	Stage (Pre-, Intra-, Post- Emergent event)	Associated Manifestation	Generating Process/Mechanism
Technically inadequate decision resulting in emergent failure in attempting to cap the well	Negative – Unintended – Unabsorbed	Intra	Single-step capping attempt failed due to technical issues related to different fluid pressures	Attempting to cement the multiple hydrocarbon and brine zones encountered in the deepest part of the well in a single operational step, despite the fact that these zones had markedly different fluid pressures
Ignition of gas released into the mobile offshore drilling unit	Negative – Unintended – Unabsorbed	Intra	The gas released into the unit was ignited without warning to operators of the pending disaster due to high volumes of gas	Various alarms and safety systems on the Deepwater Horizon Mobile Offshore Drilling Unit failed to operate as intended, potentially affecting the time available for personnel to evacuate
Disaster permitted to emerge unconstrained by regulatory controls, processes, or procedures	Negative – Unintended – Unabsorbed	Pre, Intra	Disaster emerged as technical manifestation due to underlying inadequacies in controls, regulations, and decision-making	Lack of a suitable approach for anticipating and managing the inherent risks, uncertainties, and dangers associated with deepwater drilling operations and a failure to learn from previous near misses

TABLE 9.7
Summary of Findings and Recommendations in the DHD

Topic	Needs	Total Number of Recommendations	Emergence Response Area Coverage
Improving the safety of offshore operations	New approach to risk assessment and management	3	Process
	New, independent agency	2	Learning - redesign
Safeguarding the environment	Revise and strengthen NEPA policies and practices in the offshore drilling context	1	Procedure adjustment
	Greater interagency consultation	1	System development
Strengthening oil-spill response, planning, and capacity	Improved oil-spill response planning	1	Protocol adjustment
	New approach to handling spills of national significance	1	System development
	Strengthen state and local involvement	1	System development
	Increased research and development to improve spill response	1	Learning - redesign
	New regulations to govern the use of dispersants	1	Protocol adjustment
	Need to re-evaluate the use of offshore barrier berms in spill response	1	Protocol adjustment
Advancing well-containment capabilities	Government to develop greater source-control expertise	1	Design for robustness
	Strengthen industry's spill preparedness	1	Design for robustness
	Improved capability to develop accurate flow rate estimates	1	Scanning
	More robust well design and approval process	1	Procedure adjustment
Overcoming the impacts of the deepwater horizon spill and restoring the gulf	Improved understanding of oil-spill impacts, particularly in the deepwater environment	1	Learning – propagation
	Fair, transparent compensatory restoration based on natural resource damage assessments	1	Classification – assessment
	Address human health impacts, especially among response workers and in affected communities	1	System development
	Restore consumer confidence	1	Identity development
	Long-term restoration effort that is well funded, scientifically grounded, and responsive to regional needs and public input	2	System development

(Continued)

TABLE 9.7 (*Continued*)
Summary of Findings and Recommendations in the DHD

Topic	Needs	Total Number of Recommendations	Emergence Response Area Coverage
	Better tools to balance economic and environmental interests in the gulf	1	Design for robustness
Ensuring financial responsibility	Increase existing limitations on responsible party liability	1	Protocol adjustment
	Increase limitations on payments from the oil-spill liability trust fund	1	System development
	Better auditing and monitoring of risk	1	Protocol adjustment
	Assessment of the existing claims process	1	Protocol assessment
Promoting congressional engagement to ensure responsible offshore drilling	Congressional awareness and engagement	1	System development
	Adequate funding for safety oversight and environmental review	1	System development

From the investigation, and recommendations, stemming from the DHD, we can make several conclusions. First, the response recommendations were far removed from the actual experience of emergence conditions encountered in real time for the DHD. The recommendations were not targeted, other than indirectly, to address issues of intra-emergence. Instead, the recommendations were targeted to preclude similar episodes in the future. As such, the recommendations were general in nature and nonspecific for the specific incident. Most of the recommendations were targeted to increased oversight, regulatory capacity, and accountability for planning, execution, and development. This is not intended to disparage the work of the investigation. However, in dealing with emergence, we can surmise that, while the 'macro' view is important, so too is the 'micro' view to fully understand the local, as well as global, implications for emergence. One additional point related to emergence response is appropriate. The more distant the emergent event, event context, and specific details are from the examination, the more difficult that generation of specific (local level) response becomes.

9.6 CONCLUSIONS

Emergence is a difficult concept to embrace. Not only are there multiple perspectives on what constitutes emergence, but also what might be done in response. Through the exploration of the DHD, there are several themes that have been drawn out. These themes provide a set of insights for the continuing dialog on emergence. As the field of emergence continues to evolve, we anticipate that increased developmental efforts will provide enhanced practical capabilities for more effectively dealing with emergence. Below we articulate what has been learned with respect to emergence through this exploration.

9.6.1 EMERGENCE CHARACTERISTICS CAN BE CLASSIFIED BY A ROBUST CLASSIFICATION SCHEMA

During the exploration of emergence, we discovered that developing an expanded classification schema was helpful. An expanded classification of emergence allowed for a more rigorous explanation and treatment of emergence phenomena. The inclusion of boundaries (spatial, temporal, and physical) aided in establishment of the locus of emergence. Additional elements of emergence classification included existing categories of weak/strong and intended/unintended. From a systems perspective, we added the categorization of absorbed/unabsorbed. Also, other essential attributes surrounding emergence included: location (geography), impacts (present/future), entities/system(s), time horizon, escalation, resources, processes/protocols, socio-technical-economic, controls (policies, procedures, processes), resources, internal/external manifestation, information, products, services, and functions. This heightened classification for emergence gives insight into the locus and contributing factors concerning experienced emergence. We note that emergence is generally described in very high-level terms of identification. While the classification schema is not presented as 'complete', it appears to be beyond existing classification schemas, which appear to be more superficial classifications of emergence. The more rigorous classification of emergence is an important step forward in better understanding the attributes that define an emergent event.

Design for systems to be better capable of dealing with emergence should have a focus on robustness, resilience, viability, and antifragility. The seeds for emergence can be sown well before the manifestation of an emergent condition/event. There cannot be a precise prediction of when, or in what specific form, emergence will be experienced. However, what is known with assurance is that emergence will occur in complex systems. There are four design, execution, and development aspects to the efficacy of response to emergence. First, every complex system design has a degree of robustness. Robustness of design is indicative of the range of 'potential' emergent conditions (perturbations) for which a system is capable of accommodating. While a system might not be designed for 'the specific' emergence case, it can be designed to withstand 'classes' of emergence, irrespective of the specific form emergence takes within that class. Coupled with robustness, resilience of design is the degree to which a system can return to a prior state of performance after experiencing an emergent (perturbation) condition/event. Thus, while emergence cannot be known precisely in advance, a system can be designed to better withstand and respond to a range of 'classes' of emergence. These classes of emergence may emanate from internal or external sources. A third system attribute related to emergence is design and execution for viability. Viability is concerned with assurance of continued system existence. It is an engineered attribute for complex systems with the intent of absorbing variety (emanating from the environment, or internal to a system) as emergence. Maintenance of viability is dependent on addressing emergence. A fourth attribute is antifragility. Antifragility is concerned with system modification, in response to emergence, such that the system thrives in the face of perturbations (emergence). Antifragility entails making modifications to system design or execution to compensate for experienced emergence in ways that enhance capabilities to weather future perturbation challenges.

9.6.2 EMERGENCE DOES NOT EXIST IN ISOLATION FROM THE SYSTEM CONTEXT WITHIN WHICH IT OCCURS

System context (circumstances, factors, conditions, trends, patterns) sets the conditions under which emergence occurs. Emergence in a complex system does not occur independent of the context within which the system is embedded. Context influences, and is influenced by, a complex system as well as the emergence that occurs within that context. Context can enable or constrain a complex system and contributes to the occurrence, processing, and impacts of emergent events/situations. Context is often associated with 'soft' aspects of a system (e.g., culture, management style, politics, staff, etc.). This is in contrast to 'hard' system aspects which are more technical in nature (e.g., technology, procedure, processes, etc.). While emergence is frequently identified as a hard system (technical) issue, it is remiss to remove the soft system (contextual) considerations that produce, or contribute to, emergence.

9.6.3 EMERGENCE IMPACT(S) CAN ESCALATE OR DIMINISH AS THE SITUATION AND CONCATENATION OF SUBSEQUENT EVENTS AND RESPONSES UNFOLDS

The impact(s) stemming from an emergent condition unfolds over time and the response(s) to the emergence. An 'effective' response can diminish the negative impacts, or amplify the positive impacts, of emergence. In addition, emergence does not evolve in isolation from other sequencing of information and corresponding decisions, actions, and interpretations that exist in close proximity (special, temporal, physical) to emergent events. Thus, emergence is not static but rather continues to dynamically evolve from inception of the precipitating event(s). The nature of the evolution from the initial emergent event is only partially dependent on the event. The interface of the system and system context with the emergent event will impact the trajectory of the impacts of emergence. It is naïve to consider an emergent event, and its trajectory, as being static. The preparation of a system to withstand emergent event classes attests to the potential for steering emergence to limit negative consequences and amplify positive aspects of emergence.

9.6.4 EMERGENCE PROGRESSES THROUGH A 'LIFE CYCLE' OF THREE PHASES

Every emergent event moves through three phases of a life cycle, including pre-emergence (before experiencing emergence), intra-emergence (as emergence occurs and continues to evolve), and post-emergence (in the aftermath of emergence).

Pre-emergence is focused on how a complex system is prepared for different classes of emergence (design for robustness, resilience, viability, antifragility). Intra-emergence is concerned with responses in the midst of the emergence as the event and corresponding impacts unfold. Finally, post-emergence is concerned with the aftermath of emergence and its impacts. This can range from system design modifications to changes in future execution modifications based on post-event analysis and response. It is also important to note that as an emergent event progresses, so too does 'clarity' on the nature of emergence. During pre-emergence, without prior knowledge of impending emergence specifics, the best that can be achieved is the preparation of design (structure, processes, procedures) capable of responding to 'classes' of emergence. Irrespective of pre-emergence design, not every class/variant of emergence can be accounted for. This does not diminish the need to consider pre-emergence in system design and execution planning, but rather acknowledges the limitations of resources that can be dedicated to 'emergence hardening' of a system. During the intra-emergence stage, emphasis is rightly focused on dealing with the conditions at hand and trying to place the trajectory on a course that will sustain viability as the first order of priority. The final stage of emergence, post-emergence, is where a more deliberate accounting of emergence and the adequacy of system design and execution provided for addressing emergence. It is shortsighted to simply 'survive the emergent event' and not engage in deep introspection and examination of the system in light of the emergence experienced. It is also noteworthy that post-emergence can be the most resource intensive and lengthy of the three stages.

9.6.5   EMERGENCE ENCOMPASSES A HOLISTIC RANGE OF ENABLING AND CONSTRAINING FACTORS

Emergence does not exist independently or mutually exclusive of a focal system or its context. There is a holistic range of emergent events and responses to those events that crosses the spectrum of socio-technical-economic-political dimensions. Thus, it is naïve to only consider the singular emergent event as having a limited scope (e.g., technical) and duration (e.g., instantaneous). Instead, the holistic nature of emergence suggests that the totality of the event factors should be considered across a time horizon for the evolution of the emergent event. The overly narrow consideration of emergent events misses the opportunity to engage in system development such that the system is more capable of managing 'same/similar class' emergent events in the future.

9.6.6   AMBIGUITY IS A DEFINING CHARACTERISTIC OF EMERGENCE

The nature of emergence embodies high levels of ambiguity. Ambiguity suggests there is limited clarity in understanding cause-effect relationships surrounding an emergent event, the system, and the system context. We might characterize this as the associated 'fog of emergence'. Ambiguity resolves over time as additional knowledge develops, the emergent situation evolves, and understanding of emergence comes into better focus. This does not suggest that emergence can be completely understood by analysis either during or 'after the event'. On the contrary, there can be aspects of emergence that may defy explanation and will never be completely understood. However, this does not absolve us of working to better understand and deal with 'classes of emergence' through proactive development of system designs better able to withstand emergence.

The response to emergence (e.g., DHD) is easily targeted to more high-level considerations for recommendations (e.g., broad policy implications). Unfortunately, such high-level analyses are quickly removed from the operational realities of where emergence starts and must be embraced. While higher-level global considerations of emergence are necessary, alone they are not sufficient to provide operational level analysis, assessment, and guidance to address local level systems tasked with dealing with emergence. Further development of emergence, to influence practice, must provide operational level exploration and response to emergence. For instance, more detailed classification schemas can be helpful in navigating pre-, intra-, and post-emergence activities for complex systems. The recommendations stemming from the DHD exemplify the overly broad focus that quickly morphed from the initial precipitating disaster event(s).

Emergence has always been a defining characteristic of complexity and complex systems. As the emergence field continues to evolve, so too must the focus on bringing practical utility to the design, execution, and development of complex systems. This should be a central theme in the continuing evolution of the emergence field. As the emergence field continues to evolve, we suggest three critical considerations that should help define the focus. First, the increasing robustness of emergent event classification should be expanded. What has been presented in this chapter provides more depth than what has previously been suggested. However, there is much work that must be done to enhance our capabilities and effectiveness in the classification of emergent events. Second, emphasis must be placed on bringing emergence to the realities of operational systems that experience emergence. Additional emergence development that is void of operational considerations for application is shortsighted. The development of tools, methods, protocols, and procedures to effectively deal with emergence must be a major emphasis as the field moves forward. Finally, the 'measurement' of emergence is an area that, until addressed, will represent a stumbling block for emergence in complex systems. This does not suggest that there exist a 'single set' of measures of emergence that is waiting to be discovered. On the contrary, the exploration of measurement of emergence in complex systems, the enabling conditions forecasting emergence, and the measurement of response effectiveness are all measurement considerations that should evolve as the field matures.

REFERENCES

Ackoff, R. 1994. Systems thinking and thinking systems. *System Dynamics Review* 10(2–3): 175–188.

Adams, K. M., Hester, P. T., Bradley, J. M., Meyers, T. J., & Keating, C. B. 2014. Systems theory as the foundation for understanding systems. *Systems Engineering*, 17(1), 112–123.

Aristotle. 2002. *Metaphysics: Book H - form and being at work*. (J. Sachs, Trans.) (2nd ed.). Santa Fe, CA: Green Lion Press.

Arshinov, V., & Fuchs, C. (Eds.). 2003. *Causality, emergence, self-organisation*. Moscow: NIA-Priroda.

Ashby, W. 1957. *An introduction to cybernetics*. London: Chapman & Hall, Ltd.

Ashby, W. R. 1962. Principles of the self-organizing system. In H. von Foerster & G. Zopf (Eds.), *Principles of Self-Organization* (pp. 255–278). New York, NY: Pergamon Press.

Ashby, W. R. 1991. Requisite variety and its implications for the control of complex systems. In *Facets of systems science*, pp. 405–417. Boston, MA: Springer.

Beer, S. 1979. *The heart of the enterprise.* New York: John Wiley and Sons.

Beer, S. 1985. *Diagnosing the system for organizations.* Chichester: John Wiley & Sons Inc.

Bianchi, T., Santos, D. S., & Felizardo, K. R. 2015. Quality attributes of systems-of-systems: A systematic literature review. In *2015 IEEE/ACM 3rd International Workshop on Software Engineering for Systems-of-Systems* (pp. 23–30). IEEE.

Chalmers, D. J. 1990. *Thoughts on emergence.* Online http://consc.net/notes/emergence.html, accessed 11/1/2021.

Chalmers, D. J. 2006. Strong and weak emergence. In Clayton, P., & Davies, P. (Eds.). *The re-emergence of emergence* (pp. 244–256). Oxford, Oxford University Press.

Checkland, P. 1999. Systems thinking. In *Rethinking management information systems*, pp. 45–56. Oxford: Oxford University Press

Corning, P. A. 2002. The re-emergence of "emergence": A venerable concept in search of a theory. *Complexity*, 7(6): 18–30.

Dahmann, J., & Baldwin, K. 2008. Understanding the current state of US defense systems of systems and the implications for systems engineering. In *Montreal, Canada: IEEE Systems Conference, 7–10 April.*

De Wolf, T., & Holvoet, T. 2004. Emergence versus self-organisation: Different concepts but promising when combined. In *International Workshop on Engineering Self-Organising Applications* (pp. 1–15). Berlin, Heidelberg: Springer.

El-Hani, C. N. & Pereira, A. M. 1999. Understanding biological causation. In V. G. Hardcastle (Ed.), *Where Biology Meets Psychology: Philosophical Essays* (pp. 333–356). Cambridge, MA: MIT Press (a Bradford Book).

Flood, R. L., & Carson, E. 1993. *Dealing with complexity: An introduction to the theory and application of systems science.* Berlin: Springer Science & Business Media.

Fromm, J. 2005. Types and forms of emergence. arXiv preprint nlin/0506028.

Goldstein, J. 1999. Emergence as a construct: History and issues. *Emergence*, 1(1), 49–72.

Gorod, A., Gandhi, S., White, B., Ireland, V., & Sauser, B. 2014. Modern history of system of systems, enterprises, and complex systems. In *Case studies in system of systems, enterprise systems, and complex systems engineering*, pp. 3–32. Boca Raton, FL: CRC Press

Gorod, A., Sauser, B., & Boardman, J. 2008. System-of-systems engineering management: A review of modern history and a path forward. *IEEE Systems Journal*, 2(4): 484–499.

Graham, B., W. K. Reilly, F. Beinecke, D. F. Boesch, T. D. Garcia, C. A. Murray, & F. Ulmer. 2011. The national commission on the bp deepwater horizon oil spill and offshore drilling. In *Deep water. The gulf oil disaster and the future of offshore drilling. Report to the president.* Washington: Government Printing Office.

Hitchins, D. K. 2008. *Systems engineering: A 21st century systems methodology.* Hoboken: John Wiley & Sons.

Holland, J. 1998. *Emergence: From chaos to order.* Oxford: Oxford University Press.

Holman, P. 2010. *Engaging emergence: Turning upheaval into opportunity.* Oakland: Berrett-Koehler Publishers.

Jaradat, R. M., Keating, C. B., & Bradley, J. M. 2014. A histogram analysis for system of systems. *International Journal of System of Systems Engineering*, 5(3), 193–227.

Kalantari, S., Nazemi, E., & Masoumi, B. 2020. Emergence phenomena in self-organizing systems: a systematic literature review of concepts, researches, and future prospects. *Journal of Organizational Computing and Electronic Commerce*, 30(3), 224–265.

Karcanias, N., & Hessami, A. G. 2011. System of systems and emergence part 1: Principles and framework. In *2011 Fourth International Conference on Emerging Trends in Engineering & Technology* (pp. 27–32). IEEE.

Katina, P. F. 2016. Systems theory as a foundation for discovery of pathologies for complex system problem formulation. In A. J. Masys (Ed.), *Applications of systems thinking and soft operations research in managing complexity*, pp. 227–267. New York City: Springer International Publishing.

Keating, C.B. 2005. Research foundations for system of systems engineering. In *IEEE International Conference on Systems, Man, and Cybernetics, Waikoloa, Hawaii*, pp. 2720–2725, October 2005.

Keating, C. B. 2009. Emergence in system of systems. In M. Jamshidi (Ed.), *System of systems engineering* (pp. 169–190). Hoboken, NJ: John Wiley & Sons, Inc.

Keating, C.B. 2014. Governance implications for meeting challenges in the system of systems engineering field. In *IEEE System of Systems Engineering (SOSE), 2014 9th International Conference*, pp. 154–159.

Keating, C. B., & Bradley, J. M. 2015. Complex system governance reference model. *International Journal of System of Systems Engineering*, 6(1–2), 33–52.

Keating, C.B., & Katina, P.F. 2011. Systems of systems engineering: prospects and challenges for the emerging field. *International Journal of System of Systems Engineering*, 2(2): 234–256.

Keating, C. B., & Katina, P. F. 2012. Prevalence of pathologies in systems of systems. *International Journal of System of Systems Engineering*, 3(3–4), 243–267.

Keating, C. B., & Katina, P. F. 2016. Complex system governance development: a first generation methodology. *International Journal of System of Systems Engineering*, 7(1–3), 43–74.

Keating, C. B., & Katina, P. F. 2018. Emergence in the context of system of systems. In L. B. Rainey, & M. Jamshidi (Eds.), *Engineering emergence*, pp. 491–522. Boca Raton, FL: CRC Press.

Keating, C. B., & Katina, P. F. 2019. Complex system governance: Concept, utility, and challenges. *Systems Research and Behavioral Science*, 36(5): 687–705.

Keating, C. B., & Katina, P. F. 2020. Enterprise governance toolset. In A. Gorod, L. Hallo, V. Ireland, & I. Gunawan (Eds.), *Evolving toolbox for complex project management*, pp. 153–181. Boca Raton, FL: CRC Press.

Keating, C. B., Katina, P. F., & Bradley, J. M. 2014. Complex system governance: Concept, challenges, and emerging research. *International Journal of System of Systems Engineering* 5(3): 263–288.

Keating, C., Rogers, R., Unal, R., Dryer, D., Sousa-Poza, A., Safford, R., Peterson, W., & Rabadi, G. 2003. System of systems engineering. *Engineering Management Journal* 15(3): 36–45.

Kim, J. 1997. Supervenience, emergence, and realization in the philosophy of mind. *Mindscapes: Philosophy, Science, and the Mind*, 5: 271.

Lewes, G. H. 1875. *On actors and the art of acting* (Vol. 1533). London: Smith, Elder.

Lloyds Register Foundation. 2015. *Foresight Review of Resilience Engineering*. London: Lloyds Register Foundation.

Lucas, C., & Milov, Y. 1997. *Conflicts as emergent phenomena of complexity*. Calresco Group. http://www.calresco.org/group/conflict.htm.

Maier, M. 1999. Architecting principles for systems-of-systems. *Systems Engineering*, 1(4): 267–284.

Martin-Breen, P., & Anderies, J. M. 2011. *Resilience: A literature review* (p. 64). New York: The Rockefeller Foundation. Retrieved from http://www.rockefellerfoundation.org/blog/resilience-literature-review

Murray, J. A. H., Little, W., Onions, C. T., & Friedrichsen, G. W. S. 1977. *The Shorter Oxford English dictionary on historical principles* (3rd (Thumb index) ed., completely reset / with etymologies revised by G.W.S. Friedrichsen, and with revised addenda). Oxford: Clarendon.

National Academy of Engineering and National Research Council. 2010. *Interim report on causes of the deepwater horizon oil rig blowout and ways to prevent such events*. Washington, DC: The National Academies.

Norris, F. H., Stevens, S. P., Pfefferbaum, B., Wyche, K. F., & Pfefferbaum, R. L. 2008. Community resilience as a metaphor, theory, set of capacities, and strategy for disaster readiness. *American Journal of Community Psychology*, 41(1–2), 127–150.

Plowman, D. A., Baker, L. T., Beck, T. E., Kulkarni, M., Solansky, S. T., & Travis, D. V. 2007. Radical change accidentally: The emergence and amplification of small change. *Academy of Management Journal*, 50(3), 515–543.

Rittel, H., & Webber, M. 1973. Dilemmas in a general theory of planning. *Policy Sciences*, 4(2): 155–169.

Sage, A., & Cuppan, C. 2001. On the systems engineering and management of systems of systems and federations of systems. *Information, Knowledge, Systems Management*, 2(4), 325–345.

Sauser, B., Boardman, J., & Gorod, A. 2009. System of systems management. In: M. Jamshidi (Ed.), *System of systems engineering: innovations for the 21st century*, pp. 191–217. New York City: Wiley.

Skyttner, L. 2005. *General systems theory*. Danvers, MA: World Scientific Publishing Co. Pte. Ltd.

Sousa-Poza, A. A., Kovacic, S., & Keating, C. B. 2008. System of systems engineering: An emerging multidiscipline. *International Journal of System of Systems Engineering*, 1(1/2), 1–17.

Taleb, N. N. 2012. *Antifragile: Things that gain from disorder*. Vol. 3. New York: Random House Incorporated.

US DoD (Department of Defense). 2008. Systems Engineering Guide for System-of-Systems. 05–04.

10 Emergent Behavior in Space Architecting and On-orbit Space Operations

Larry B. Rainey

Integrity Systems and Solutions of Colorado, LLC

CONTENTS

10.1 CHAPTER INTRODUCTION

This chapter has two sections. The first is a reprint of the article *Space Architecting: Seeking the Darwinian Optimum Approach* written by Dr. Roberta M. Ewart, Chief Scientist of the Space and Missile Center at Los Angeles, CA. In essence, she addresses emergent behavior in space architecting from an acquisition planning perspective. She gave this paper at the American Institute of Aeronautics and Astronautics (AIAA) SPACE 2011 Conference and Exposition on 27–29 September 2011 at Long Beach, California. I obtained approval from Dr. Ewart on 14 November 2020 in an email from her to reprint her article. Secondly, Dr. Rainey addresses the reality of how emergent behavior can be found in on-orbit operational system of systems.

DOI: 10.1201/9781003160816-12

10.2   DR. ROBERTA EWART'S ARTICLE

The last few years have seen an explosion in quality decision support tools and decision support frameworks providing analytic foundations for architecting system of systems and families of systems. Systems engineers have been eagerly asking for such a capability in order to integrate an ever more complex community of space systems across the National Security Space (NSS). For Air Force Space Command, a new concern for architecting is how to seamlessly include the cyber architecting strategies from this point forward. To date, no overarching or guiding theme united these architecting efforts and their toolsets and provided overall guidance to the space design community. In addition, without some architectural guidance, it has been very difficult determining the science and technology (S&T) investment strategies as well as the concurrent acquisition strategies for such a complex situation. Succinctly, if an architecting construct or framework could be devised, it would have benefits across many aspects of the NSS, including architecting, S&T management, requirements generation and procurement.

One approach for addressing the architecting framework, and the answering the architecting shortfall, is what the author call's the Darwinian Optimum Approach, or, DAO. The DAO built out of a decision support framework of many tools, has several features that are detailed in this paper. These features spring from the skill sets of systems engineering, operations research analysis, and science and technology management; hence, the framework which employs them will require diverse teams to successfully generate products. The DAO uses biological analogues and basic Darwinian theories mixed with modern analytic tools optimized by operations research and systems engineering principles. For example, the tools can be used to operate on highly diverse options in the early phases of concept development (i.e., growth), followed by a strong down selection mechanism to a dominant design or a set of features in a given niche (e.g., Analysis of Alternatives). This dominant or optimum design is known as the government reference design (GRD). The GRD can then be put into other niches for further analysis of adaptability. The down selection in any niche is based on measurable qualities, similar to the qualities of successful Darwinian candidates. Similarly, the DAO *evolves* over time, based on the inherent characteristics of the architecture (both hereditary and non-hereditary traits) that provide the most benefit (e.g., have higher fitness). The DAO must deal with complexity, at least on the order of the complexity of the operation of the human body and the anthropomorphic or biological emphasis serves as framework to extend the overall research strategy into the cyber realm. In this way, numerous medical and immunological applications can quickly be adapted for use in the architecting process for future space cyber systems.

This paper is aimed at addressing how to organize analytic tooling to instantiate a unified architecting methodology and possibly unified space architecture. This paper outlines an approach to begin building a unified analytic tool set and architecting to Darwinian optimum standards.

10.2.1 Background

NSS has no overarching architecture methodology nor integrated architecture. Current space systems are built as unique programmatic efforts as point designs with some ability to integrate after the systems are instantiated. In the past, this has not been a major problem for acquires, who find it easier to deliver single (MILSATCOM, Weather, Missile Warning (MW), Position Navigation and Timing, etc.) families, with long-lived programmatic budget lines, rather than integrated portfolios (COM/NAV, Weather/NAV, MW/NAV, etc.), crossing traditional budget lines, to the warfighter. However, the warfighter has begun to change the paradigm. The warfighter now wants to instantiate a service-oriented functional delivery of capability which can be tailored at the operational, strategic, and tactical level to allow the tailoring of functionality and provide more flexibility Given this predilection for the service-oriented approach presented at the operational level, it is now critical to instantiate an architecting methodology to guide space enterprise development.

There are two additional background issues which affect the reception of this paper.

First, there are numerous definitions of the term "architecture" which can make any discussion of architecture-related topics difficult. For the purposes of this paper, the architecting methodology is aimed at multi-perspective decision support for use across the NSS enterprise. This methodology is designed primarily for force modernization planning and is designed to be conducted in concert with systems engineering (SE) management and control, and program integration. Due to its multi-perspective nature, it can also support CONOPS development, planning, programming and budgeting analysis, and requirements generation and refinement. In the simplest terms, "this architecting methodology is designed to be analytic and create knowledge or courses of action for decision makers, as compared to other methods which document existing baselines and are intended primarily for configuration management purposes or making very minor material changes to the architecture."

Second, "DAO" is not technically an acronym. DAO is a variant for TAO which is as known as "The Way of Heaven" in Chinese philosophy. The "Way" is achieved when the individual (each component) and society (the entire architecture) operate in the right manner.[1] It is intended as a way to focus the attention of the reader onto the holistic nature of architecting inspired by the Darwinian approach. It is also a means of highlighting the important balance of Eastern (DAO) and Western (Darwinian) traditional approaches when building frameworks for dealing with complexity. Using the diversity inherent in these two approaches, it is more likely the framework derived from the two viewpoints can successfully solve the very difficult problem of architecting the space enterprise.

10.2.2 CONTEXT FOR DARWINIAN ARCHITECTING

The National Security Space Strategy (NSSS)[2] Hs highlighted the key features of the space environment, and these changes call for commensurate changes in the space enterprise design environment. In brief, space is viewed as congested, contested, and competitive environment. These three features are inherent in other military force structure analysis processes but are relatively new to the space community when taken together. Each of these three features points to the need to modify the overall architecting context. In addition, the space and cyber engineering communities are coming together at many levels to collaborate on how to deliver future space cyber systems. The engineers are looking at systems from a physical layer represented by networks, up to the highest layer of applications or "APPS." This implies space cyber integrated designs should investigated at these multiple layers and their interfaces. By taking three features and the constraint that the architecting must now be done for the space cyber systems, the stage is set for proposing Darwinian-based architecting.

First, addressing the contested feature of the space cyber systems, systems engineers are often not required to design space systems to face the full spectrum of threats in the contested environment. Since the end of the Cold War, many space systems have had vulnerability or survivability requirements "sold off" to lower bid price in order to afford the basic capability the system could deliver. This is also stated as: "the system obtained a waiver to key performance parameters for survivability and vulnerability." This waiver approach implies those space systems are probably not robust in the current contested environment. This robustness shortfall is occurring at the same time the nation is relying more on its space systems and the asymmetric benefits they have delivered to our deployed forces.[3] To remedy this shortfall, or at a minimum characterize the shortfall, anyone creating an architecting methodology for NSS should consider returning to "Opposed System Design" as envisioned by Albert Wohlstetter.[4]

Opposed systems design has many analogous, one of which is the approach Darwin documented when he described how species interact in nature to survive and whose memorable catchphrase is "survival of the fittest." Opposed systems design implies the engineer will have to look at all levels of the opposition the systems will face, from natural to manmade actions, both intentional and unintentional. Though the Darwinian approach did not describe behaviors to the level of political context, i.e., the Cold War, as the Wohlstetter approach did, nor deal with "intentioned" human behavior, the features of his analysis are very useful for helping frame the analysis needed to make space systems robust in a contested environment. In sum, a Darwinian approach can give insights and solutions for architecting systems in the contested environment which might otherwise not occur to the designers.

Second, the US space industrial base is no longer competitive at of levels, both domestic and international. Other competitors are offering diverse goods and services which demonstrate better perceived value for many niches, and the US is no longer the dominant space provider. (This is not the fault of the industrial base, but an outcome of many interactions, values, and policies pursued over a long period of time.) There is low diversity at all levels of the US industrial base supplying the components for space systems. Industry prime contractors often decide the major outcomes of the NSS competition within their design space which leads to artificial versus natural competition and artificial down selection for NSS procurements. However, the NSS procurement system is based on federal acquisition principles which require a level playing field and encourage the growth of competitive forces. This is succinctly stated in policy terms in the NSSS, page 13, "We [US] seek to address *competition* by enhancing our own capabilities, improving our acquisition processes, fostering a healthy U.S. industrial base, and strengthening collaboration and cooperation." So again, any architecting methodology must help illuminate how to close the existing lack of natural competition by offering diversity of candidates and understanding of the competitive niches. This need for diversity points to a need to study the Darwinian forces seen in nature and propose how they might be applied to the study of the space enterprise.

Third, space is now physically congested by satellites owned by more actors (nation states and corporations), and an ever-growing debris population. Using the Darwinian analog, it is clear that "niches," originally sparse with competitors, and with fixed resources for the competitors, are being squeezed by the addition of players. Using the Darwinian context, it is possible to study the competition and determine what key traits a successful competitor has obtained, convert those into metrics, and then optimize them to design more robust and adaptable space cyber systems. In sum, the space cyber systems of today must be designed within a framework which recognizes congested, contested, and competitive features. The proposed Darwinian approach is designed to use opposed systems design, principles of diversity, and observation of successful multi-niche behaviors related to performance metrics successful competitors display during down selection. DAO, if exercised across multiple niches, can result in multiple courses of action for enterprise design decisions and can provide a new perspective while utilizing much of the existing workforce and their skill sets. This approach has emerged as the one most likely to be useful to decision makers who "often find the analytical results and final recommended solution less important than identifying and articulating the problem, as well as organizing the information associated with it."[5]

10.2.3 Existing Disciplines Supporting Architecting Methods

To start the process of improving decision support and instantiating a DAO, it is necessary to characterize the current workforce and their approaches to architecting, set a future vision, and then propose how to close the gap between the two. This section of the paper characterizes the existing disciplines.

There are three primary disciplines currently contributing to architecting processes and products and which also represent the bulk of the workforce and the talents for doing architecting. The first discipline, operations research (OR) is defined as "a scientific method of providing executive departments with a quantitative basis for decisions regarding the operations under their control."[6] In the Darwinian optimum approach, OR will be used to link desired performance traits to systems requirements traces and is represented by the term "phenotype." OR practitioners would need to adjust to the DAO by studying biological populations and key interactions among populations. This would include predator to prey interaction, food/water, and procreation competitions. Many of these areas already have mathematical representations common in format to mathematical representations used for studying military operations or supply chains. The analyst would need to study both equation-based (deterministic and stochastic processes) and agent-based or object-oriented environments, which are currently being explored for solving emergent behaviors of large populations of individuals.[7] The survey of this existing body of knowledge should uncover guides for long- and short-lived trends representing techniques deemed to be successful in the niche. It is likely that the analyst currently using either set of methods could be called upon to set up a verification and validation methodology to check for the work of the other type of analysis. In other words, the OR contributions to DAO should be able to both generate phenotypes and check phenotype behaviors regardless of their predilection for agent or equation-based methods. There are already discussions on the balance of agent-based and equation-based approaches for doing this type of analysis.[8] One notable strength of an agent-based approach is that if the agents represent proprietary designs, it may be possible to set up the analysis so that it is necessary to see the internal workings of the agent. This feature may allow multiple parties to play the phenotypes in the trade space without jeopardizing their future competitive position of revealing their "genotypes." Genotypes are associated with the SE approach which is discussed next.

SE is a discipline which links requirements traces to functional components and eventually to physical instantiations of objects. In the DAO, the systems engineers will need to learn how to convert from a parametric and equation-based approach, as represented by Excel spreadsheets, MATHCAD, and Mathematica (all well-known commercial tools which the author does not specifically endorse) which are used across the whole community, to also being comfortable with agent-based approaches. Systems engineers also need to share a set of performance measures understandable to the OR community in order to translate between the phenotype behaviors and the genotype instantiations of specific systems representing the competing species. But based on existing work, the phenotype performance measures of accessibility, availability, capacity, coverage, delay, latency, reliability, scalability, speed, survivability, and timeliness already exist. Additional work would need to be done to create appropriate metrics for adaptability, autonomy, flexibility, and robustness. The author proposes that by starting with the first set of performance measures, operating the DAO to see emergent behaviors, and checking those emergent behaviors with equation-based verification and validation approaches, the SE team members in DAO would be able to arrive at appropriate metrics for the latter set. To date, the key research from which to build this type of capability would be from "value-centric design methodologies" applied to NSS systems engineering problems.[9] The Defense Advanced Research Projects Agency (DARPA) is also investing in efforts to help quantify adaptability, autonomy, flexibility, and robustness within its F6 program efforts. Specifically, DARPA has stated that "the first technical area [in the BAA] is aimed at the development of design tools that appropriately quantify the adaptability of a design, alongside its performance, cost, and risk. The goal is to appropriately incentivize the incorporation of adaptability features into space system designs by facilitating analytically rigorous design trades between adaptability and traditional system attributes at each level of abstraction in the design."[10]

SE will be generating species, laying out the constraints of genotyping representing those species, and OR will analyze the species behaviors, feeding back that information to the SE. One near-term example would be to analyze the functional bloodlines with genetic linage/traces with desired behaviors of operationally responsive space (ORS). The author has already placed OR and SE disciplines into a similar relationship by using the DoD modeling and simulation system, where four M&S layers are built into an iterative and recursive architecting methodology and conducted architecting analysis. The top two layers of the analysis framework, campaign, and mission modeling are primarily operated by the OR community, and the lower two layers—the engagement and engineering layer—would primarily be populated by the SE tools. See Figure 10.1.

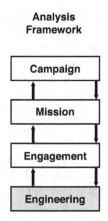

FIGURE 10.1 DoD modeling and simulation hierarchy for architecting.

The third and final "discipline" supporting DAO may not be considered a "discipline" itself, but Decision Theory (DT) represents a key skill set for successfully linking the OR and SE disciplines to any useful outcomes. DT basically links analysis to force structure decisions. DT as practiced within DAO requires both the decision processes and the decisions themselves to be modified for both short- and long-term survival scenarios and fully integrated with the phenotypes and genotypes. DT supplies the knowledge to shape the visual tooling and social representation for the processes and products in the DAO. The DT baseline for DAO is drawn from RAND authors, Lempert, Popper and Banks research and their theory of robust decision making (RDM).[11] The RAND team has designed four basic steps or rules for RDM which DAO can implement freely.

The first rule is consider "large" numbers of scenarios. In the DAO, a reasonable number is six to eight scenarios. These scenarios are generally available through DoD modeling, and simulation organizations or vignettes based on the scenarios can be generated in "reasonable" amounts of time, normally days to a week. The scenarios need to represent "plausible futures" that are diverse as possible. In the case of DAO, the plausible futures will contain extrapolations of the current space enterprise threat environment which includes natural radiation and man-made threats such as directed energy, kinetic, and cyber and include the current enterprise systems such as communication, navigation, weather, and sensing spacecraft as the DAO genotypes. The second rule will be to "seek robust", rather than optimal strategies that do "well enough" across scenarios. The robustness feature will be quantified as an emergent metric, but one starting point might be the system must be successful at meeting their performance metrics a certain percentage of the time in all six to eight scenarios. This percentage could initially be set at the 50% level and then examined at the 80% level and even beyond.

The third rule is employ adaptive strategies. This sounds very pedantic but would analyst would have to see emergent behaviors which are different over the short term. The long-term successful behaviors are slowly varying characteristics which probably provide a survival advantage over many decision cycles while the short-term behaviors would be those used only in one decision cycle. One example of this would be the fuel consumption on a spacecraft. In the long run, survival is based on very carefully managing and as much as possible limiting fuel usage. However, under a short-term threat scenario, a spacecraft may have to decide to expend precious fuel to adjust to the threat. The rule behavior for making the short- and long-term decisions will be related but different. A short-term decision ultimately adjust the long-term decision as the fuel is consumed. This behavior is exhibited with predator cats in Africa and the prey herd animals that interact with them. The predator must carefully weigh how quickly it thinks it can get the prey before it has hit the energy expenditure rule set limiting its chase. A decision cycle could be modified after a single observe, orient, decide, and act loop (OODA) commonly used in DoD OR. Fourth, RDM recommends the use of computer tools designed for "interactively exploring multiple plausible futures." The interactive nature of the fourth rule is closely linked to the idea from the OR discipline which points the importance of the decision maker who will use the products to have a hand shaping the products. Having now described the three foundational "disciplines" upon which to build the DAO, the next section discusses how to initialize the DAO for space architecting.

10.2.4   POSSIBLE PATH TO ESTABLISH DAO

Briefly, the description of how to instantiate this methodology is as follows. The envisioned architecting methodology will be built out of an analytic decision support framework of OR, SE, and DT tools, each of which has the ability to assess the key metrics for service-oriented functional delivery. These metrics would be drawn from a pool or taxonomy of terms such as accessibility, adaptability, autonomy, availability, capacity, coverage, delay, latency, reliability, robustness, scalability, speed, survivability, and timeliness. To date, three metrics have already been examined for space situational awareness: capacity, coverage, and timeliness; this analysis has created an excellent foundation upon which to extend the DAO effort. The DAO toolset will be built from experience gained by analyst in traditional disciplines, such as astronautical, mechanical, electrical engineering, military, and commercial OR and decision support theory, which implies the DAO cadre needs to probably start with approximately nine personnel, probably almost equally allocated across the three disciplines. This would equate to approximately an initial investment of $3 M assuming a fulltime would run about 0.3 M. The SE tools will be used as their underlying and unifying driver but hybridized with modern agent–oriented analytic tools. The agent-oriented tooling will most likely be built on the System Effectiveness Analysis Simulation (SEAS),[12] a modeling and simulation environment already in the USAF Modeling and Simulation portfolio and maintained by SMC Development Planning, XR.

The "optimization" component of the effort shall focus on applications related to complexity and chaos theories which allow local areas or surface of solutions in the multi-dimensional trade space or attractors in the trade space to pivot toward emergent behaviors desired by the designers or showing where the dominant design has advantages or disadvantages in a given niche. Initially, the DAO will be used to operate on highly diverse options in the early phases of concept development (i.e., growth), followed by a strong down selection mechanism to a dominant design or set of features in a given niche. This dominant or optimum design will be known as the government reference design (GRD). The GRD can then be put into other niches for further analysis of its adaptability. The GRD may also represent a government knowledge point and act as a baseline to assess any future system offered by contractor in a bid to then build the GRD. The contractor would be required to match or improve on the performance of the GRD under the same constraints. This implies that at some point industry would be knowledgeable of the DAO process and be required to run their own versions with fidelity as good or better than that which generated the GRD. The DAO will be aimed to link key developmental planning components: (1) analysis of future threats, strategy, and needs; (2) advanced concept engineering; (3) capability analysis and gap identification; and (4) course of action creation to deliver decision quality products to decision authorities.

In sum, the down selection effects in any niche, and with varying fidelity will be based on measurable qualities, similar to the qualities of successful Darwinian candidates, where "phenotypes" represent the qualities and "genotypes" represent the instantiated systems design, but down selection must be linked back to service-oriented metrics. So rather than food or procreation competitions as biological systems, the DAO will emphasize competition for capacity, coverage, and timeliness of the space cyber system performing a function. The DAO will *evolve* over time, based on the inherent characteristics of the architecture where both hereditary and non-hereditary traits are contemplated that provide the most benefit (e.g., have higher fitness). Often the hereditary traits serve as the slow variants of a system similar to the attractors or stable surfaces in chaos/complexity theory studies, while the non-hereditary traits are new outliers that can assist in finding other stability points or attractors that could have not otherwise been found but which act over shorter timeframes.

The DAO must deal with complexity and related chaos theory, at least on the order of the complexity of the operation of the human body, and the anthropomorphic and biological emphasis will serve as a framework to extend the overall research strategy. In this way, numerous medical and immunological applications can quickly be adapted for use in the architecting process or offer paradigms for new genotypes (material solutions) to insert into the competitive environment. To date, the cyber community has found biological analogues very useful for their analysis, and this acceptance should speed the acceptance of this approach for space cyber systems. Some examples of cyber analogues for space cyber which are drawn from the biological realm are the creation of functional "skin" which can help ward off broad attacks against the space cyber entity and the understanding of antibodies and autoimmune responses. Current exciting research applications include the use of statistical physics approaches to study antibody responses.[13] Space systems designers can leverage existing cyber analysis by extending the capability into space systems design by assuming space systems are extended nodes of cyber system, and all space systems are "compromised" systems and have inherent disease vectors design flaws.

The DAO is composed of heterogeneous tools and must be able to represent multiple perspectives of the instantiated material solutions. The minimum number of perspectives includes the preexisting space systems engineering framework which uses the engineering "V." See Figure 10.2.

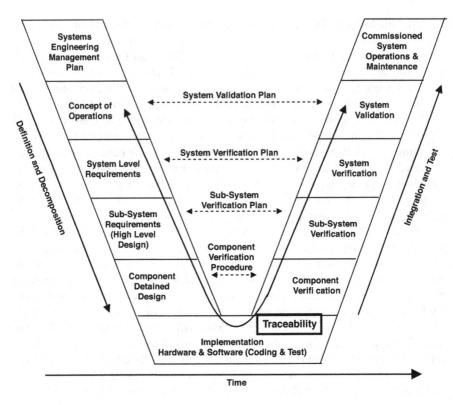

FIGURE 10.2   The standard systems engineering "V."[14]

Cyber has only recently developed a utility space linked to information transfer and has had an open architecture, ad hoc design approach, typically employing a network layer model composed of seven sub-perspectives called the Open Systems Interconnection (OSI) layer model to manage its development. See Figure 10.3.

The DAO will force the community to delve deeper into the seven layers to reconcile the space and utility functions using "service" performance parameters used in the cyber world. The DAO will help the space cyber system engineering community to balance the traditional space systems engineering discipline with the open systems ad hoc approach of cyber. In addition, the DAO will reconcile the OR and SE perspectives to provide a relevant portfolio to the warfighter.

The DAO is already being used to guide the initial outline of the space cyber S and T portfolio.[15] The key features of the space cyber S and T portfolio link to three emphasis areas or research lines in complex systems, autonomy, and emergent behavior. These three emphasis areas will have crucial impacts on the nature of the OODA loop and could represent the first breakthroughs associated with using the DAO. Improvements in the OODA mimic the natural evolution of the neural-based capability in the many species. The biological equivalent of the observe phase has generated multiple phenomenologies based on multiple senses in many species. In the space cyber case, a new "sense," the infrared sense, has been added to the observe phase

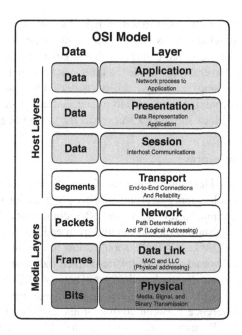

FIGURE 10.3 Internet stack perspective of space cyber systems.[16]

of the OODA. The orient phase is an extension of the evolution which has created holistic neural frameworks with rapid pattern recognition. Pattern recognition has led to a learning capability and the associate skills for predicting possible outcomes which gave the human species its first edge in survival. Accurate prediction capabilities continue to be the edge in any conflict and are usually associated with leadership and command experience that comes with age. Improvements in the orient phase are also one area where the addition of cyber capabilities could show the greatest improvements and leverage both pattern recognition and accurate prediction skills.

Autonomy can be represented by the ability of organisms to locally "fight through" and survive in their niche without the help of human intervention. The decide phase will ultimately benefit from improve autonomy as decision authority will be placed at the optimal place in the man-machine interfaces. One area of additional autonomy analysis is how to adjust the decision phase of the OODA so that it is more integrated into the orient phase and begins in the OODA cycle. One proposition is that the cyber analysis of orientation will be the decisive edge to push this research further. Current discussion is often focused on the difference between "automation" of the lower-level tasks man is not well suited for autonomy reflecting the man to machine boundary by becoming the virtual partner in many space systems, and robotic exploration has allowed man even greater opportunities to experiment with the spectrum of automation autonomy. DAO-based analysis can take this research even further.

Emergent behavior should be one of the greatest legacies of employing the DAO. The author predicts that behaviors will emerge from the analysis which preferentially favors design heuristics for robustness and flexibility which ultimately provides adaptability. However, the DAO should ensure that the design heuristics do not emerge from artificially constrained niches. Assuming early superiority in a niche guarantees long-term success needs to be revised. Designers must now assume rapid niche change for the space enterprise. Some emergent behaviors to immediately address are maneuverability (inside the kinetic act enemy loop), responsiveness (inside the act enemy loop), reconfigurability (better OODA act components), and graceful degradation (passing learned info to succeeding generation and retaining some capability in all phase of the OODA even as the system begins its demise).

10.2.5 DAO Pathfinders

It is now useful to consider some examples of how the Darwinian optimum architecting approaches could supply decision support products for upcoming modernization or recapitalization issues facing the NSS community. Two examples are considered: satellite operations (SATOPS) and overhead persistent infrared (OPIR) systems. Applying the DAO to SATOPS should allow immediate space cyber heuristics to emerge which favor advanced software-defined radios or cognitive radios and software-enabled geodesic dome-phased arrays. These two technologies can be used to develop the concept of the AF Cyber gateway, which assumes there is no topological or operational difference between command and control of spacecraft and the data they produce. OPIR analysis allows operational difference between command and control of spacecraft and data they produce. OPIR analysis allows early exploration of multiple niches and transition from phenotype to genotype. The GRD sensors can be played in multiple scenarios based on the niches established by battlespace awareness, technical intelligence, missile warning, and missile defense to establish the desired characteristics of the phenotype. In this case, sensitivity, coverage, and timeliness might be more desired phenotype behaviors. It would then be up to DAO to determine optimum genotypes for obtaining IR-based information for multiple niches (BA, TI, MW, and MD) at LEO, MEO, GEO, and HEO by iterating such things as monolithic IR systems, rideshare smaller IR satellites, secondary payloads, and possibly hosted payloads and their constellation designs. The possibilities for DAO are quite large.

10.2.6 Summary

There is a need to develop a space cyber architecting framework. One possible approach is to use a Darwinian construct of survival of the fittest as a means of designing complex systems capable of thriving in the congested, contested, and competitive environment of space. Using tool sets developed for OR, SE and DT, it is possible to develop this architecting framework to assess flexibility and robustness and related adaptability of complex systems. It may be possible to analyze the recent "resilience" construct as part of an ongoing space cyber architecting effort.

10.2.7 Conclusion

In order to improve the current NSS enterprise modernization trade space, it is necessary to create more capable decision support processes with richer options. In order to get this diversity, new perspectives from non-traditional or external disciplines and viewpoints must be considered. One proposed option is to use the Darwinian approaches from the study of biology as a means to reframe the key architecting methodology for architecting the space enterprise. This method could reconcile the key analytic approaches already used in the space community with the new approaches being pursued by the cyber community to generate space cyber architectures. Additionally, the DAO would bring together several key disciplines supporting advanced decision analysis. Along with its sister paper, the author hopes that this paper will encourage further discussion on the nationally critical topic and guide the future investment strategies using a disciplined methodology that can accept and use emergent behaviors of advanced technologies—particularly to support key decision processes and augment human decision making. The ultimate goal shall be an NSS architecture that becomes a standards-based, net-centric, service-oriented virtual enterprise.

10.3 EMERGENT BEHAVIOR IN SPACE ARCHITECTING AND OPERATIONS: DR. RAINEY'S COMMENTS

In the above article by Dr. Ewart, there are actually two levels of analysis with respect to the subject of emergent behavior in system of systems. They are planning/acquisition and operational. The latter is only mentioned once by Dr. Ewart in her above article. Line two of her above abstract states "support frameworks providing analytic foundations for architecting system of systems ...."[17] In Maier's article entitled Architecting Principles for System of Systems, he states that there are "Five principal characteristics that are useful in distinguishing very large and complex but monolithic systems from true system of systems."[18] They are "Operational Independence of the Elements, Managerial Independence of the Elements, Evolutionary Development, Geographic Distribution and Emergent Behavior."[18] He defines emergent behavior as "The system performs functions and carries out purposes that do not reside in any component system. These behaviors are emergent properties of the entire system-of-systems and cannot be localized to any component system. The principal purposes are the system of systems are fulfilled by these behaviors."[18] Continuing the discussion Chapter 1 identifies that agent-based modeling can be used to test for the presence of emergent behavior. Next Chapter 1 identifies the two types of emergence, i.e., positive and negative, and provides definitions of each. It should also be noted, for the reader, that Dr. Ewart also addresses this former subject on multiple occasions in a companion article.[19]

The second level of analysis, i.e., planning/acquisition, is most definitely addressed by Dr. Ewart in detail. She states "...it is now critical to instantiate an architecting methodology to guide space enterprise development."[17] Her approach is the "Darwinian Optimum Approach, DAO."[17] She goes on to state

> It is intended as a way to focus the attention of the reader onto the holistic nature of architecting inspired by the Darwinian approach. It is also a means of highlighting the important balance of Eastern (DAO) and Western (Darwinian) traditional approaches when building frameworks for dealing with complexity. Using the diversity inherent in these two approaches it is more likely the framework derived from the two view points can successfully solve the very difficult problem of architecting the space enterprise.[17]

Furthermore, she states "There are three primary disciplines currently contributing to architecting processes and products and which also represent the bulk of the workforce and the talents for doing architecting."[17] They are OR, SE, and DT. For the former, she states

> In the Darwinian optimum approach, OR will be used to link desired performance traits to system requirements traces and is represented by the term "phenotype". OR practitioners would need to adjust to the DAO by studying biological populations and key interactions amongst populations. [17]

For the middle discipline, she states

> SE is a discipline which links requirements traces to functional components and eventually to physical instantiations of objects. In the DAO the systems engineers will need to learn how to convert from a parametric and equation based approach, as represented by Excel spreadsheets, MATHCAD and Mathematica (all well known commercial tools which the author does not specifically endorse) which are use across the community, to also being comfortable with agent-based approaches.[17]

She states that

> The third and final "discipline" supporting DAO, may not be considered a "discipline" itself but Decision Theory (DT) represents a key skill set for successfully linking the OR and SE disciplines to any useful outcomes. DT basically links analysis to force structure decisions. DT as practiced in DAO requires both the decision processes and the decisions themselves to be modified for both short and long term survival scenarios and fully integrated with phenotypes and genotypes. DT supplies the knowledge to shape the visual tooling and the social representation for the processes and products in the DAO.[17]

Note from the above that Chapter 1 identifies that agent-based modeling can be used to test for the presence of emergent behavior. How applicable for Dr. Ewart to also endorse agent-based methods for the same purpose. She states "The analyst would need to study both equation based (deterministic and stochastic processes) as well as **agent-based** (emphasis added) or object oriented environments which are currently explored for solving emergent behaviors of large populations of individuals."[3] She goes on to state: "In other words, the OR contributors to phenotype should be able to both generate phenotypes and check phenotype behaviors regardless of their predilection for agent or equation based methods. There are discussions on the balance of agent based and equation based approaches for doing this type of analysis. One notable strength of an agent based approach is that if the agents represent proprietary designs, it may be possible to set up the analysis so that it is not necessary to see the internal workings of the agent."[3] Later in her article, she states "The agent oriented tooling will be most likely be built on the Systems Effectiveness Analysis Simulation, SEAS, a modeling and simulation environment already in the USAF Modeling and Simulation portfolio...."[17] The author knows from personal experience that SEAS is an agent-based tool.

In Chapter 1, Dr. Rainey provided definitions for both positive and negative emergent behavior as well as addressed the subject of desirable and undesirable outcomes. Dr. Ewart in her DAO article appears to address this theme. She states

The "optimization" component of the effort shall focus on applications related to complexity and chaos theories which allow local areas or surfaces of solutions in the multi-dimensional trade space or attractors in the trade space to point towards emergent behaviors **desired** (emphasis added) by the designers or showing where the dominant design has advantages or disadvantages in a given niche.[3]

Later in her article, she states

Emergent behavior should be one of the greatest legacies of employing the DAO. The author predicts that behaviors will emerge from analysis which preferentially favors design heuristics for robustness, and flexibility which ultimately provide flexibility. However, the DAO should ensure that the design heuristics do not emerge from artificially constrained niches.[3]

In addition, in this article she states "Applying the DAO to satellite operations (SATOPS) should allow immediate space cyber heuristics to emerge which favor advanced software defined radios or cognitive radios and software enabled geodesic dome phased arrays."[17] In her companion article, she states

The Darwinian evolution and adaptation techniques can be linked together in such a way to allow for emergent behaviors to become more obvious to decision makers so that the command and control systems could evolve at faster rates as well.[19]

Clearly, all of the above are DAO oriented for space architecting purposes.

10.4   LESSONS LEARNED

As stated above, there are actually two levels of analysis with respect to the subject of emergent behavior in system of systems. They are operational and planning/acquisition. The third article above by Dr. Ewart addressed her "architecting methodology to guide space enterprise development." As such, her article clearly addressed the latter level of analysis. Interesting enough, however, she initiated her article by stating "The last few years have seen an explosion in quality decision support tools and decision support frameworks providing analytic foundations for architecting **system of systems and families of systems** (emphasis added)." However, she did not go into detail as to describe "decision support tools and decision support frameworks" for operational system of systems.

It is her above second later companion article that is also focused on the planning/acquisition level of analysis yet it also has definitive implications for the former level of analysis, i.e., operational. In her introduction to her above second later companion article she stated

It has become clear over the past few years that the science and technology execution plans for space and cyber systems were on an intersection path. That intersecting path has resulted in a new paradigm called space cyber. Through collaborative efforts amongst government, industry and academia partners not only can the space cyber paradigm be addressed, a resilient space cyber S and T strategy can be proposed. In order to meet this need, this paper proposes a four step execution process for resilient space cyber systems.

It is in her first and third steps of her execution plan where specific reference is made to the above operational level of analysis. In her first step section, she states Figure 10.1. AF Cyber Vision 2025 Approach

> embodies the fulfillment of NSS S & T resilient cyber execution plan's first step by guiding a collaborative environment within government and industry and begins the process of recapitalizing legacy spacecraft designs and **system of systems** (emphasis added) with newer and more resilient options.

In her third step, she states "...future engineering tasks will require efficiently coordinated distributed parallel and stochastic processing jobs to devise large distributed architectures **system of systems and families of systems** (emphasis added)."

It is also in her S and T Research Lines for Resiliency section where there are three references to the operational level. First, she states "The remaining recommendations are those S and T strategies which could be iterated in the testbeds or implemented in a more robust and flexible networked satellite **operations system of systems** (emphasis added)." Second, later she states "The research lines for collective behavior, nonlinear dynamics and pattern formation also probably offer the best departure point for the situational awareness components of the current **system of systems** (emphasis added) used by the warfighter." Finally, she states

> ..the implementation of the S and T line would be to take an agent based environment such as System Effectiveness Analysis Simulation (SEAS), embed key attractors which represent **current systems or system of systems** (emphasis added) and then do a stability analysis by iterating amongst several niches (scenarios).

From a lessons-learned perspective, the conclusion that can be drawn is that Dr. Ewart's primary focus has been on the programmatic/acquisition aspects for space systems versus on the actual operational aspects of space systems development itself. She addressed the subject of emergent behavior in the context of her Darwinian Optimum Approach for space system development/acquisition and her efforts for resilient space cyber systems. The bottom line lessons learned are twofold. First, emergent behavior as defined by Maier in Section 10.2 applies to operational systems. Second, as addressed by Dr. Rainey in Section 10.2, definitions for both positive and negative emergent behavior were provided in Chapter 1 as well as addressed the subject of desirable and undesirable outcomes. This latter perspective by Dr. Rainey has been addressed by Dr. Kristin Giammarco in her online briefing entitled "Practical Modeling Concepts for Engineering Emergence in System of Systems" (https://sercuarc.org/wp-content/uploads/2018/07/2018-03-20-SoSECIE-Giammarco-brief.pdf). On Slide 12, she states "Positive emergence is what remains after thoroughly exposing and removing Negative emergence." Slide 13 provides a five-step algorithm to eliminate negative behavior. This procedure is very important as negative emergence could potentially significantly impair the mission of a system of systems. Conversely, positive emergence could potentially be a force multiplier for a system of systems.

The above conclusion is extremely pertinent to national defense. "Under a project called Space Vision 2030, members of the consortium will be asked regularly to submit ideas and concepts for future national space architecture."[20] "We have to modernize the space architecture into what we think we'll need 10 years from now,"[20] said Major Vinney Pande, satellite prototyping program manager at the Space and Missile Systems Center. In a related document, it was stated "Some will pose large **system-of-systems** (emphasis added) concepts, while others may be tailored to specific components or sensors as enabling technologies."[21]

10.5 CHAPTER SUMMARY

This chapter has been an examination of the above article by Dr. Roberta Ewart addressing emergent behavior in space system architecting. Dr. Ewart's article was provided for the reader's context. Next, Dr. Rainey provided his comments on her article. These pertained to her programmatic/acquisition perspective but also addressed the operational perspective. This chapter also addressed Lessons Learned.

REFERENCES

1. Edwards, P., (ed), *The Encyclopedia of Philosophy*, Macmillan Publishing Inc, New York, 1967. vol. 2, pg 88.
2. National Security Space Strategy: https://www.dni.gov/reports/201 national security space strategy.pdf
3. National Space Policy, 28 June 2010, https://www.whitehouse.gov/sites/default/files/national_space_policy_6-28-10.pdf
4. Wohlstetter, A., "Theory of Opposed Systems Design", D(L)-16001-1, August 1967 (revised January 1968).
5. Loerch, A. G., and Rainey, L. B., (ed.), "Introduction to Operational Analysis", *Methods for Conducting Military Operations Analysis*, Military Operational Research Society, MORS, 2007, pg 4.
6. Morse, P. M., and Kimball, G. E. *Methods of Operations Research*, MIT Press, Cambridge, MA, 1951.
7. Wilson, W.G. "Resolving Discrepancies Between Deterministic Simulation Models and Individual Based Simulations", *American Naturalist*, 1998 Feb; 15(2), pgs 116–34.
8. Parunak, H. V. D., Savit, R., "Agent Based Modeling vs Equation Based Modeling: A Case Study and Users Guide", Proceedings of Multi-Agent Systems and Agent-Based Simulation (MABS '98, Paris) 10–25.
9. Ewart, R., Betser, J., Gasster, S., Gee, J., Penn, J., Richardson, G., "Enhancing NSS Systems Engineering Analysis Through Value-Centric Design Methodology", Proceedings AIAA SPACE 2009, Pasadena, CA, September 14–17, 2009.
10. System F6 Draft Broad Agency Announcement (BAA), Solicitation Number: DARPA-SN-11-01, Oct 2010, pg 5, https://wwwfbo.gov/index
11. Lemper, R., Popper, S., "Robust Decision Making: Coping With Uncertainty", *The Futurist*, Feb, 2010, vol. 44, no. 1. https://www.wfs.org/J-F2010/Jan FB BKis.htm
12. SEAS information and downloads can be found at: http://teamseas.com
13. Gaidos, S., "Physicist Join the Immune Fight", *Science News*, Jan 15, 2011, pgs 22–25. http://www.sciencenews.org
14. Figure 2 can be found at: http://:commons.wikimedia.org/wiki/File:Systems Engineering V diagram.jpg
15. Ewart, R., et al., "Space Cyber Technology Horizons", SPACE 2011, American Institute of Aeronautics and Astronautics, Reston, VA. (submitted for publication).
16. Figure 3 can be found at: http://en.wikipedia.org/wiki/OSI model

17. Ewart, R., "Space Architecting: Seeking the Darwinian Optimum Approach", AIAA SPACE 2011 Conference & Exposition, 27–29 September 2011, Long Beach, California.
18. Maier, M. https://www.deepdyve.com/lp/wiley/architecting-principles-for-systems-of-systems-pz4RlAjpyP
19. Ewart, R., "Executing A National Security Space Science and Technology Strategy for Resilient Space Cyber Systems", AIAA SPACE 2012 Conference & Exposition, 11–13 September 2012, Pasadena, California.
20. https://spacenews.com/on-national-security-space-force-to-industry-lets-talk-future-plans/
21. Introduction to Space Vision 2030, Space & Missile Center information document.

Section III

Summary

Summary

11 Lessons Learned and the Proposed Way-Ahead

Larry B. Rainey
Integrity Systems and Solutions of Colorado, LLC

O. Thomas Holland
Georgia Tech Research Institute

CONTENTS

11.1 LESSONS LEARNED

11.1.1 CHAPTER 2

- System of systems are a type of complex system.
- Emergence occurs in system of systems.

DOI: 10.1201/9781003160816-14

- Emergence in system of systems can be positive when it is by design and useful to the purpose of the system of systems.
- Similarly, emergence in system of systems can be negative when it is unexpected and especially when it interferes with the purpose of the system of systems.
- Emergence can be increased by increasing the variety available to the entities comprising a system.
- Emergence can be reduced by increasing the constraint (decreasing the variety) on entities.
- Simulation is a means to explore emergence in system of systems.
- New modeling techniques allow us to discover scenarios that produce emergent behaviors.

11.1.2   CHAPTER 3

- While all authors understood the comprehensive scenario generation capability of Monterey Phoenix (MP), some gravitated toward using MP as they would employ a traditional tool and understood the benefits to be limited to automatic generation of more scenarios. Some were not used to modeling using abstractions and were at first tempted to model every implementation detail. This led to some initial concerns about the scalability of MP for systems with many (perhaps billions) of nodes, but with coaching it was realized that modeling at full scale was not necessary to obtain useful models and insights – that only small examples are needed to make inferences that also apply at scale. This is the Small Scope Hypothesis in action, and seeing it work on examples relevant to each author helped to solidify understanding of how to use MP. The amount of MP model coaching required by the different authors varied – some authors were able to develop a model independently after a two-hour tutorial, and then get insights from the model quite quickly with no coaching at all; others had more complex models requiring the formal expression of non-trivial behaviors benefitting from having access to a modeling coach. None of the participating authors dismissed MP for cosmetic differences from familiar tools and, as a result, were able to learn and leverage its unique value proposition of exhaustive behavior scenario generation for small scopes, finding unexpected emergent behaviors permitted by their designs.

11.1.3   CHAPTER 4

- Emergence happens everywhere.
- Dealing with emergence (both positive and negative) has to be seen by a human to understand its impact.
- Past models for understanding outcomes in a war at sea scenario make assumptions that are helpful but not always accurate.
- Pairing of Monterey Phoenix and Lanchester equations and the Hughes salvo model could create a new level of understanding.

- Emergence includes the systems, as they are combined in system of systems, context, and the role of dynamics.
- In warfare, there are risks that 'understanding' by humans within the system will be inadequate, forcing a reliance on alternative means.
- The OODA loop is an important framework from which to develop a Monterey Phoenix model from a scenario.

11.1.4 CHAPTER 5

- Cybersecurity threats can be characterized with the terminology of complex negative emergent behaviors that are difficult to control using existing processes and technologies.
- Block Frame, Inc., has demonstrated unique processes and technologies that address current cybersecurity vulnerabilities using a system of systems suite to provide defense-in-depth and zero-trust architectures using novel technologies, including cryptologic provisioning devices, highly scalable registries, and indefinitely scalable blockchains/distributed ledgers to manage trust and security through visible dashboards and controls.
- Model-based systems engineering can be used to characterize and quantify the effects of both negative and positive emergent behaviors in the cybersecurity environment. The Naval Postgraduate School Monterey Phoenix Model can be used to describe high-level functional behaviors of normal operations, operations under attack, and processes to mitigate cyber threats.

11.1.5 CHAPTER 6

- **A Wargame Is Interactively Complex**: A wargame can be considered as a system of systems in which the component parts of objectives, research questions, design, methodologies, adjudication protocols, etc. operate in a predictable way to produce wargame outcomes. As such, a wargame may be considered to be structurally complex in that it produces an orderly operation achieving emergent outcomes. But, there is an additional component which completely changes the nature and operating principles of a wargame... human participants considering a complex, dynamic, and operational environment. These participants and their comprehension of an evolving situation completely alter the characteristics of a wargame and transform it into an interactively complex structure which operates with a freedom of action to produce unpredictable emergent behavior.
- **Emergent Behavior Rests upon the Human Intellect's Ability to Separate, Emphasize, and Eliminate in the Consideration of a Problem**: Emergence in wargaming is not the result of a simple relationship or interaction among components. Instead, it relies upon the intellect's desire for understanding and resolution. This desire causes the participants to engage the game field and impose order and coalescence upon fluid variables that, through the action of state changes as the game advances, allows insights and direction to form which constitute the emergent outcomes.

- **Emergent Behavior Is Actually a Sum of Constituent Emergences Which Occur During Game Play**: Wargaming rests upon too many interacting variables and fluid situations to permit the assembly of a final emergent behavior without intermediate emergent compositions. Emergence in wargaming is a combination of sub-emergences whose values, emphases, and relationships will change with the evolution of game states. Narrative emergence is the 'story' the participants build in a game as it progresses. It is vital as a means of establishing connection, coherence, and purpose as the game states evolve and the scenario develops. Dynamic emergence is the change in the complex operating environment under the duress of decisions and interactions which cause permutations to influence the value, emphasis, and relationships among selected areas of action. Intermediate emergence is the local production of adjudication results that occur to influence the formulation of a new game state. As this process is continuous in the life of a game, a cumulative emergence will be generated as the sum of and final expression of the combination of narrative, dynamic, and intermediate emergence in the definition of and assessment of the final game state.
- **The Cumulative Emergence of a Wargame Contains the Summation of the Wargame's Emergent Behavior Which Resides in the Game's Final State and Which Is Extracted with and Given Substance to by a Human Cognitive and Computational Evaluation Process**: The final state of a wargame is represented by the expression presented in Section 6.2 and below. Here the actions of the speed of thought upon cognitive calculation and consensus and the maneuver of knowledge upon informed judgment and decision to produce emergence are associated so that, combined with adjudication and a resulting evaluation, the current state of a wargame is defined. The sum of these continued and repeated actions eventually produces a final game state such that εc is the cumulative (final emergence); SN is the final game state, (eN, eD, eI) are the narrative, dynamic, and intermediate emergences; and E is the final analysis, evaluation, and assessment procedure.

$$\varepsilon_c = S_N \left(e_N, e_D, e_I \right) \cdot E$$

- **Monterey Phoenix Provides Useful Support to the Phases of the Wargame Process**: Monterey Phoenix is a high-level language that allows for behaviors to be expressed and executed. The ability to generate behavior scenarios from combinations of perceived requirements provides a means to expose and control process behaviors, including those that are emergent. This is applicable in wargame preparation, execution, and assessment. In wargaming preparation, wargame designers can inject emergent behavior opportunities in the behaviors of game components and so emphasize intended consequences or deemphasize unintended consequences depending on the intent of the wargame. In the context of wargaming execution, the ability to view and explore behaviors in the form of annotated event traces provides an intuitive means for verifying and validating potential courses of action, along with their excursions to better identify suitable actions. The same capabilities enable modeling and analysis of wargame solutions to identify and delineate possible excursions for subsequent wargames, i.e., reuse of wargaming behavioral constructs.

11.1.6 CHAPTER 7

- Continuous monitoring and controlling mechanisms must be enforced to prevent unintended behaviors of Internet of Things (IoT) devices. These mechanisms are based on the normal behaviors of IoT devices within an organization. Any abnormal behavior should trigger an alarm to the administrator to investigate the cause of the abnormality.
- Correct modeling of IoT devices should be performed using their data profiles. Modeling tools are then used to investigate different scenarios and interactions that can arise between system components. Additionally, simulating the IoT system using simulation tools provides insights for system administrators on the amount of data traffic that can be expected based on the number of devices within the system.
- Billions of unprotected and uncontrolled devices can be used to perform unprecedented scenarios. These devices are exploited by bad actors (hackers) to perform illegal actions. Recent IoT botnets are real examples where botmasters control many devices and send them commands to perform attacks such as DDoS attacks with unprecedented traffic volume. Modeling and simulation can reveal such scenarios where proactive solutions and mechanism can be deployed to mitigate threats.

11.1.7 CHAPTER 8

- The structure of the neural network is key to the presence of emergence, and this structure must form organically, itself an example of emergence on a longer time-scale than the shorter dynamical time scale.
- Neurons and the synapses that interconnect them have a number of regulatory processes that allow for interesting collective behaviors to emerge. These include:
  - Excitatory and inhibitory neurons. In the examples studied here, what we would identify as the emergent behavior is observed primarily in the excitatory neurons, but the inhibitory neurons are needed to stabilize the behavior.
  - Auto-regulatory behaviors that vary the responsiveness of a neuron to inputs in order to prevent saturation.
  - Regulatory processes for the weakening and strengthening of synapses (network plasticity) to limit the rate of network change, maintain certain network features, and/or prevent degeneracies.
- These processes and the motifs in the way they are employed could be valuable in understanding potential designs for emergence in other system of systems, even those that are not explicitly neuro-inspired.

- Graph signal processing is a developing tool for analyzing dynamics and interactions on networks, but the peculiarities of biological neural networks (directed and signed) are an under-studied area.
- The proper choice of Graph Fourier Transform can identify low-dimensional structure that is characteristic of emergent behavior and can also identify functional components in a neural network.
- The combination of graphical and standard Fourier analysis, leading to a Joint Vertex-Time transform, can further reveal this low-dimensional behavior by detecting behaviors that evolve both temporally and spatially across the network.

11.1.8 CHAPTER 9

- Emergence characteristics can be classified by a robust typology. During the exploration of emergence, we discovered that developing an expanded typology for classification was helpful. An expanded typology allowed for a more rigorous explanation and treatment of the phenomena. The inclusion of boundaries (spatial, temporal, and physical) aided in establishment of the locus of emergence. Additional elements of typology included existing categories of weak/strong, intended/unintended, and absorbed/unabsorbed. Also, other essential attributes surrounding emergence included: entities, control processes, resources, internal/external manifestation, information, products, services, functions, and conceptual. This heightened classification for emergence gives insight into the locus and contributing factors concerning experienced emergence.
- Design for emergence should have a dual focus on robustness and resilience. The seeds for emergence can be sown well before the manifestation of an emergent condition. There cannot be a precise prediction of when, or in what specific form, emergence will be experienced. However, what is known with assurance is that emergence will occur in complex systems. There are two design aspects to the efficacy of response to emergence. First, every complex system design has a degree of robustness. Robustness of design is indicative of the range of 'potential' emergent conditions (perturbations) for which a system is capable of accommodating. While a system might not be designed for 'the specific' emergence case, it can be designed to withstand 'classes' of emergence, irrespective of the specific form emergence takes. Coupled with robustness, resilience of design is the degree to which a system can return to a prior state of performance after experiencing an emergent (perturbation) condition. Thus, while emergence cannot be known precisely in advance, a system can be designed to better withstand and respond to a range of 'classes' of emergence. These classes of emergence may emanate from internal or external sources.
- Emergence does not exist in isolation from the system context within which it occurs. Emergence in a complex system does not occur independent of the context within which the system is embedded. Context is taken as the circumstances, factors, conditions, and trends that influence, and are

influenced by, a complex system. Context can enable or constrain a complex system and contribute to the occurrence, processing, and impacts of emergent events/situations. Context is often associated with 'soft' aspects of a system (e.g., culture, management style, politics, staff, etc.). This is in contrast to 'hard' system aspects which are more technical in nature (e.g., technology, procedure, processes, etc.). While emergence is frequently identified as a hard system (technical) issue, it is remiss to remove the soft system (contextual) considerations that produce, or contribute to, emergence.

- Emergence impact can escalate or diminish as the situation and concatenation of responses unfolds. The impact(s) stemming from an emergent condition unfolds overtime and the response(s) to the emergence. An 'effective' response can diminish the negative impacts, or amplify the positive impacts, of emergence. In addition, emergence does not evolve in isolation from other sequencing of information and corresponding decisions, actions, and interpretations. Thus, emergence is not static but rather continues to dynamically evolve from inception of the precipitating event. The nature of the evolution from the initial emergent event is only partially dependent on the event.
- Emergence progresses through a 'life cycle' of three phases. Every emergent event moves through three phases, including pre-emergence (before experiencing emergence), intra-emergence (as emergence occurs and continues to evolve), and post-emergence (in the aftermath of emergence). Pre-emergence is focused on how a complex system is prepared for different classes of emergence (design for robustness and resilience). Intra-emergence is concerned with responses in the midst of the emergence as the event and corresponding impacts unfolds. Finally, post-emergence is concerned with the aftermath of emergence and its impacts. This can range from system design modifications to changes in future execution modifications based on post-event analysis and response.
- Emergence encompasses a holistic range of enabling and constraining factors. Emergence does not exist independently or mutually exclusive of a focal system or its context. There is a holistic range of emergent events and responses to that event that crosses the spectrum of socio-technical-economic-political dimensions. Thus, it is naïve to only consider the singular emergent event as having a limited scope (e.g., technical) and duration (e.g., instantaneous). Instead, the holistic nature of emergence suggests that there the totality of the event factors should be considered across a time horizon for the evolution of the emergent event.
- Ambiguity is a defining characteristic of emergence. The nature of emergence embodies high levels of ambiguity. Ambiguity suggests there is limited clarity in understanding cause-effect relationships surrounding an emergent event. Ambiguity resolves over time as additional knowledge develops, the emergent situation evolves, and understanding of emergence comes into better focus. This does not suggest that emergence can be completely understood by analysis 'after the event'. On the contrary, there can be aspects of emergence that may defy explanation and will never be completely understood.

11.1.9 CHAPTER 10

- Dr. Roberta Ewart (Chief Scientist of the Space and Missile Center at Los Angeles Air Force Base, CA) in her article Space Architecting: Seeking the Darwinian Optimum Approach only addresses emergent behavior in the context of space architecting from an acquisition perspective. She only uses the term system of systems once without elaboration.
- Dr. Roberta Ewart (Chief Scientist of the Space and Missile Center at Los Angeles Air Force Base, CA) in her related article "Executing a National Security Space Science and Technology Strategy for Resilient Space Cyber Systems" uses the term system of systems six times without elaboration.
- Emergent behavior needs to be examined in the context of proposed on-orbit system of systems using Monterey Phoenix to first model on-orbit system of systems. Second Monterey Phoenix needs to be used to assess both the existence of negative and positive emergent behavior and to delete the negative emergence such that only positive remains. This could significantly enhance mission effectiveness of the on-orbit system of systems. Negative emergence could potentially negatively impact the mission of an on-orbit system of systems. Positive emergence could potentially be a significant force multiplier for an on-orbit system of systems.

11.2 PROPOSED WAY-AHEAD

The following texts have been developed that address emergent behavior in system of systems from an academic perspective:

1. Modeling and Simulation Support for System of Systems Engineering Applications[1]
2. Engineering Emergence: A Modeling and Simulation Approach[2]
3. Emergent Behavior in Complex Systems Engineering: A Modeling and Simulation Approach[3]

This text "Emergent Behavior in System of Systems Engineering: Real-World Applications" is the first to address real-world applications as the title suggests.

What is now required to move the discipline forward is the development of a professional organization that is analogous to the American Institute of Aeronautics and Astronautics (AIAA) and the Institute of Electrical and Electronic Engineers (IEEE). AIAA and IEEE came into existence to provide a professional discipline and a collegial forum to lead to exceptional results in each respective area. The same is true for the subject of emergent phenomena. The publication of research papers and texts is growing on a daily basis. As such, there is the need for a professional society to provide discipline and a collegial forum that will lead to exceptional results. Such a proposed organization is the Consortium for the Study and Exploitation of Emergent Phenomena (CSEEP) in complex systems. This professional organization or society does not currently exist. Rather, such an organization needs to be stood-up and formally established. CSEEP should be a 'call to arms' for the community to recognize what needs to be done to further this discipline.

The following proposal is suggested:

Toward the Creation of a Consortium for the Study and Exploitation of Emergent Phenomena in complex systems (CSEEP)

11.2.1 GUIDING PRINCIPLES

Beyond the technical understanding of emergence in complex systems, the world will benefit from a collaborative body of practitioners, researchers, manufacturers, educators, and policy makers that will help derive the benefit from emergence and avoid the pitfalls.

11.2.2 VISION

The CSEEP vision is to establish a nationally recognized collaboration of emergence practitioners, researchers, producers, policy makers, and educators that will lead business, government, and personal interests in the responsible, effective, and beneficial development and application of emergence arising from a future of complex[4] and system of systems.

11.2.3 MISSION

The mission of the CSEEP is to promote the responsible exploitation of the emergence phenomena by providing an independent, not-for-profit resource for research guidance, commercial development, general education, and public policy.

11.2.4 STRATEGY

Simply stated, CSEEP will achieve this mission by providing the necessary thought-leadership, consultation, knowledge management, collaborative infrastructure, and technical insight to:

1. Connect emergence research with product innovators.
2. Advise policymakers on the appropriate concerns related to emergence phenomena impact on society.
3. Educate the public about the realities of emergence and how to benefit from it.
4. Assure that the United States and its allies are prepared to consider emergence in system of systems and operations to assure the defense and safety of the public.

11.2.5 BACKGROUND

What is meant by Emergent Phenomena?

Emergent phenomenon, often referred to only as emergence, is a debated topic in the scientific literature pertaining to complex systems and system of systems. CSEEP adopts the definition put forth in the doctoral dissertation by O. Thomas Holland where Emergence is defined as '...those phenomena in which patterns that are observed at a global level arise solely from interactions among lower-level components acting on rules that are executed using only local information without reference to the global pattern'.[5]

11.2.6 THE NEED

It is easy to see that we are already experiencing a seismic shift in the way we perceive and interact with the world just by noting the new vernacular: terms such as 'drone attack', 'cyber-physical system', 'self-driving car', and '3D printing' are mentioned daily. Combinations of unprecedented technological advancements are dramatically affecting the way people interact with the environment and each other. These advancements include the increasing computer power brought about by ever more capable computational hardware and cloud computing; the increased ability to interact with data through virtual reality, augmented reality, and new human-computer interface technologies; the ever-increasing connectedness giving rise to the IoT; the advances in manufacturing, such as micro-miniaturization of electronics, nanotechnology, and additive manufacturing; the rapid advance of biological engineering, such as genetic manipulation and cloning; increasing richness in social structures, such as social media and the near-instantaneous dissemination of information. These and many other new discoveries and technologies still on the horizon are not just changing how we live in the environment but are changing the very nature of what we call environment. How will humanity adapt? Can humanity adapt?

Beneath the surface of all of this change is a concept called complexity. By complexity, we mean the degree, i.e., the extent, of how things interact with and depend on each other. As individual things, called entities, interact over time, we find that these interactions give rise to persistent combinations of entities we call structures that cannot be described using the same measurements used to describe the constituent entities. These new, evolving, and often unexpected structures we say 'emerge'. Holland refers to such resultant structures as emergent behavior systems (EBS), i.e., systems that consist of multiple independent yet interacting entities, evolving and reacting to changes in environment and information, sustained on the edges of chaos without any overarching or centralized control. EBS can include complex systems that are both complex structurally and dynamically, occurring either naturally or by design, as well as modern concepts of system of systems where complex systems are formed from the interaction of operationally and managerially independent systems.[6] Emergent behavior systems in and of themselves are neither good nor bad – they just are; they are the resultant of many interacting entities brought about through feedback paths and responses to stimuli. They can be intentional, as in system of systems, or unintentional when they arise from unforeseeable interactions or events. It is the result/effects of EBS that have either preferable or non-preferable outcomes.

In spite of the challenges presented by complexity, or rather because of complexity, emergent behavior systems exist all around us both in the natural and man-made world. Ant-hills grow and take in food without any individual ant knowing the grand plan. Birds migrate in great flocks. Traffic snarls. Hurricanes form. Terrorist organizations threaten the world. Economic bubbles form and burst. Still, we struggle to detect the onset of emergence and are often caught by surprise when it happens. We have even greater difficulty planning for emergence. We want to design robust and dynamic systems that can combine relatively simple entities (usually meaning easy to create and inexpensive) to achieve sophisticated goals which would be impossible by individuals. For example, we want to enable a single user to pilot a swarm of hundreds of micro-UAVs or exert subtle influences on complex situations to avert uprisings or catastrophes. Additionally, we want to use combinations of sophisticated systems, each built for specific purposes, and use them in combination with other systems to create an effect unimagined when the individual systems were conceived. Often in such system of systems applications, we treat emergence as a phenomenon to be avoided and so constrain them to prevent emergence but, in doing so, create non-adaptive (brittle) systems that underperform and often fail catastrophically. Although some inroads to understanding complexity and emergence have been made, there remains much we don't yet know how to understand.

11.2.7   What Is Meant by Exploitation?

We use the term exploitation to mean the use of emergence to purposefully achieve some worthwhile goal; that is, how can we influence entities to drive the emergent to a desirable behavior? There are many questions to explore, such as: Are there patterns of relationships between entities that suggest the tendency to certain kinds of resultants? Can we develop expressive formalisms that allow emergence to be explored in abstraction? Can a causal link between the micro-scale and the macro-scale be expressed? Since the resulting effects can be positive/advantageous or negative/not advantageous, can some agent or mechanism be specified that promotes the positive effects and manages the negative effects?

11.2.8   Legal Entity: 501–3c (Nonprofit)

CSEEP will be organized as a not-for-profit 501–3c association to foster its mission and goals. The intent of CSEEP is not to endorse nor produce for-profit products directly, but to advance the need for scientific study, education, and governance of emergent phenomena technologies and their applications by bridging researchers with producers. CSEEP will endeavor to connect researchers with producers to encourage new and innovative applications of emergent phenomena as the science and technologies are developed.

11.2.9   Next Steps: Establishing a Board of Governors

CSEEP is not a system of systems organization per se nor a complexity research institute. It is an organization that will connect researchers, product innovators, policy makers, educators, and operational planners so that there is a common understanding of emergence and its implications/uses. In essence, it is an organization to provide leadership in this area of endeavor. To accomplish its mission, CSEEP is founded on a Board of Governors comprised of thought leaders in the areas of academia, government, and industry.

11.2.10 ESTABLISHING A CSEEP'S PRESENCE

Several of those who have participated in the initial interest meetings of CSEEP are already published in the areas of emergence and systems engineering. There are already existing various systems science, systems engineering, and complexity science publication venues that exist that can be used to document CSEEP research and findings and spur continued interest.

REFERENCES

1. Larry B. Rainey and Andreas Tolk (eds). *Modeling and Simulation Support for System of Systems Engineering Applications*. John Wiley and Sons, 2015.
2. Larry B. Rainey and Mo Jamshidi (eds). *Engineering Emergence: A Modeling and Simulation Approach*. CRC Press, 2019.
3. Saikou Diallo, Saurabh Mittal, and Andreas Tolk, (eds). *Emergent Behavior in Complex Systems Engineering: A Modeling and Simulation Approach*. John Wiley & Sons, 2018.
4. In this sense, a system comprised of multiple autonomous entities that interact to produce a measurable effect.
5. O. Thomas Holland. *Partitioning Method for Emergent Behavior Systems Modeled by Agent-based Simulations*. Old Dominion University, 2012.
6. Mark W. Maier. Architecting principles for systems-of-systems. *Systems Engineering: The Journal of International Council on Systems Engineering*, 1(4), 267–284, 1998.

Index

Note: **Bold** page number refers to tables and *italics* page number refers to figures.

Taylor & Francis eBooks

www.taylorfrancis.com

A single destination for eBooks from Taylor & Francis
with increased functionality and an improved user
experience to meet the needs of our customers.

90,000+ eBooks of award-winning academic content in
Humanities, Social Science, Science, Technology, Engineering,
and Medical written by a global network of editors and authors.

TAYLOR & FRANCIS EBOOKS OFFERS:

A streamlined
experience for
our library
customers

A single point
of discovery
for all of our
eBook content

Improved
search and
discovery of
content at both
book and
chapter level

REQUEST A FREE TRIAL
support@taylorfrancis.com

Printed in the United States
by Baker & Taylor Publisher Services

Printed in the United States
by Baker & Taylor Publisher Services